Introduction to Nutrigenomics

INTRODUCTION TO NUTRIGENOMICS

Edited by:

Bhupendra Pushkar

Delve
PUBLISHING

www.delvepublishing.com

Introduction to Nutrigenomics

Bhupendra Pushkar

Delve Publishing
224 Shoreacres Road
Burlington, ON L7L 2H2
Canada
www.delvepublishing.com

Email: orders@arclereducation.com

ISBN: 978-1-77469-479-4 (Hardcover)

ABOUT THE EDITOR

Dr. Bhupendra Pushkar was born in Mumbai, India. He persuaded his B.Sc. in Microbiology in 2000. He has obtained his M.Sc. degree in Biotechnology in 2002 from University of Mumbai. He pursued his further studies and completed his Doctorate degree in 2010. His area of research broadly includes environmental pollution remediation using biological science. He is currently working on heavy metal bioremediation, PAH bioremediation, plastic degradation, plant tissue culture etc. He is actively involved in the bioremediation research work and trying to develop system for river water cleaning. Since 2008 he is working as assistant professor at University of Mumbai, India. He has published various papers in national as well as international journals and currently guiding three research students. He is Co-Editor in Chief of a journal and reviewer in many journals.

TABLE OF CONTENTS

LIST OF FIGURES

ABSTRACT

The interaction between nutrition and cell/genetic processes is known as nutrigenomics or nutritional genomics. It tries to explain the impacts of diet on a person's genes and overall wellbeing. Nutrigenomics attempts to offer a genetic understanding of how regular dietary substances (i.e., nutrition) impacts the synchrony between health and infection by modifying the expression or design of a person's genetic composition. The science analyzes the reaction of people to food elements applying post-genomic and similar technologies (for example proteomics, metabolomics, genomics, transcriptomics, and so on). Nutritional genetics is considered as the blend of nutrigenomics and nutrigenetics. This volume on nutrigenomics establishes the impacts of ingested nutrients and other food particles on genome expression and regulation. It likewise defines the personal nutritional needs based on the genome makeup of the individual (customized diet) and the relationship between diet and terminal diseases which will assist with understanding the etiologic factors of terminal conditions like a malignant growth, type-2 diabetes, obesity and heart disease.

PREFACE

Nutritional genomics is the research on interrelations between the genetic composition, food varieties we eat, and the health results that may occur. Nutritional genomics is a comparatively new study field that envelops two unique fields: Nutrigenomics and nutrigenetics. Nutrigenomics centers on how nutrients or non-nutritive bioactive elements, for example, chemical inhibitors present in food impact gene expression, protein and metabolite concentration and thus digestion, health condition and disease risk. Nutrigenomics is a field of nutritional genomics that concentrates on the impacts of food sources on gene expression. This sounds complex, however, this volume presents it in a rather simple way.

For instance, red wine has become something of a cutting-edge food thanks to its high resveratrol content. This compound stimulates the gene that shields cells from free radical damage. Another supplement that impacts gene expression is folate, present in food sources like vegetables and fruits. Folate is a nutrient required by the body to form DNA. Without sufficient folate, you have a greater risk of getting cancer. Since the common diet has numerous bioactive ingredients that can through the interrelation with receptors enact or balance the expression of target genes or directly cause the restitution of chromatin structure, it is broadly acknowledged, however not frequently perceived in the planning and translation of genetic and epidemiologic research.

The research that follows the impact of a given diet often dismisses the conceivable impact of genetic variability within the focus group, then again, some studies reviewing the impact of candidate genome quality polymorphism on the studied characteristic (for instance pulse, weight) do not comprise the diet regimen, which can drastically impact the subsequent relationship. Dietary genomics aims to resolve this obvious error.In a given parallel with pharmacogenomics, nutritional genomics centers around the bioactive substances found in regular food and how those substances influence the balance between health and infection by interacting with the person's genome.

This volume mentions that the quest for knowledge in regards to healthy/sufficient food has expanded within the last 10 years, among the global population, nutritionists, and health experts. Since early times, people have realized that climate and food can affect a person's health conditions, and apply used plants and food as medicine. With the development of science, particularly after the Human Genome Project (HGP), researchers began questioning whether the association between genes and food bioactive elements could positively or adversely affect a person's health. To evaluate this relationship between genes and nutrition, the phrase "Nutrigenomics" was developed.

Subsequently, Nutrigenomics relates to the application of nutrition, biochemistry, physiology, genomics, metabolomics, transcriptomics, epigenomics and proteomics, to review and make sense of the current proportional collaborations among genes and supplements at a sub-atomic level. The detection of these relationships (gene-nutrient) will help with the preparation of customized diets as indicated by every person's genotype. In this manner, it will be feasible to alleviate the side effects of existing conditions or to fore stall future diseases, particularly in the field of Non-transmissible Chronic Diseases or (NTCDs), which are presently viewed as a significant international public health issue.

CHAPTER 1

UNDERSTANDING NUTRITIONAL EPIDEMIOLOGY

CONTENTS

1.1 WHAT IS EPIDEMIOLOGY?

It is the investigation of how disease spreads within a community. The fact that disease does not predictably manifest itself provides a foundation for future research. As a result, it is a useful tool for investigating the causes of nutrition-related illnesses and making recommendations for prevention or therapy. Many interdependent processes are required for the proper construction and function of our cells, tissues, organs, and the entire body. At the systemic level, this is accomplished by providing the body with the nutrients it requires and allowing it to convert the substrates and cofactors required for metabolism to occur in the first place (Wiwanitkit, 2012). Physical activity levels, which can change in response to physiological and pathological illness conditions, have an impact on food intake, energy, and nutrition requirements. Diet and metabolic capacity, in other words, nutritional epidemiology encompasses not only food and beverage intake, energy and nutrient consumption, and body composition, but also indicators of nutritional status in the body and nutritionally influenced biological processes.

To decode multiple nutritional elements, multiple aspects of physical activity, such as intensity, duration, and frequency, must be evaluated. The observational nature of epidemiology can have both positive and negative consequences. Nutritional epidemiology investigates dietary and nutritional factors concerning the prevalence of illness in populations. Nutritional epidemiology is a relatively new field of study in medicine that investigates the relationship between diet and health. Although it is a relatively new field of epidemiology, it is becoming increasingly relevant to current health concerns. "Epidemiologists may be overlooking nutrition because it is more difficult to measure than other disease risk factors." Nutritional epidemiology seeks to better understand human nutrition and explain basic concepts through the use of nutritional science data (Alagawany et al., 2022). Nutritional epidemiology is based on nutritional science and includes clinical, case-control, and cohort studies in its research and therapeutics. To investigate the link between food and illness, nutritional epidemiology methods have been developed.

Figure 1.1: Epidemiology illustrations.

Source:https://www.news-medical.net/health/Statistics-and-Epidemiology.aspx

These findings have an impact on dietary recommendations, particularly those tailored to avoid specific diseases and disorders, as well as malignancies. According to Western experts, all health and social care professionals should be educated on nutritional epidemiology due to its growing importance and track record of effectiveness in improving global public health. The role of food in health and illness has been defined as exposure with significant measurement error in nutritional epidemiology studies. Nutritional epidemiology is a sub-discipline of epidemiology that studies nutrition (Wiwanitkit, 2012). It provides information on the relationship between nutrition and disease through Public Health Nutrition. This data is then used to develop preventative practices in the field. The work of nutritional epidemiology includes measuring and modeling the link between food and disease.

According to this article, there has been significant progress in all of these areas in recent years. Due to the possibility of significant measurement errors, large-scale epidemiological research is shifting away from the food frequency questionnaire (FFQ) as the primary dietary assessment instrument and toward short-term quantitative techniques. Web-based self-administration technologies are now being investigated due to the high cost of interviewers' 24-hour recalls. Furthermore, there is a strong emphasis on documenting and evaluating each meal consumed, which may improve nutritional evaluation accuracy. Food labels should be standardized across studies to improve the description of nutrient consumption and increase comparability. This new high-throughput technique will allow researchers to better understand how bioactive chemicals indicate food consumption and status by using mass spectrometry to analyze concentrations in bodily fluids as potential biomarkers.

To obtain accurate estimates of diet-disease associations, statistical modeling of dietary data and the interactions between diet and disease can employ algorithms that convert quantitative short-term evaluations into individuals' habitual intakes and data from biomarker validation studies. For nutritional data, substitution modeling is preferred over simple addition modeling. Nonlinear interactions, on the other hand, require more investigation. Due to the increasing difficulty of conducting and evaluating research in this field, young scientists entering the field of nutritional epidemiology require specialized training. As a result, the number of high-quality articles on nutritional epidemiology topics will increase in the future (Afman & Müller, 2006). In recent years, nutritional epidemiology has faced numerous criticisms, including its reliance on observational research to address etiologic concerns and its failure to adequately assess food. Recent meta-analyses with major methodological flaws have resurrected previously identified issues by producing false or misleading results.

All of these factors have cast doubt on the ability of nutritional epidemiology studies to influence public policy. These arguments are primarily based on a misinterpretation of the field's methodological limits and an incorrect application of the drug trial paradigm to nutrition research. Human food is of interest in nutritional epidemiology because it is a complex system of interacting components that have a cumulative effect on health. As a result, nutritional epidemiology is always confronted with new problems that need the creation of novel solutions. A lack of understanding of these elements can lead to ineffective and sometimes uninformed criticism. This section attempts to explain common misconceptions about nutritional epidemiology, address professional constraints, and assess the value of nutritional research in influencing policy by focusing on five major issues commonly raised about the topic.

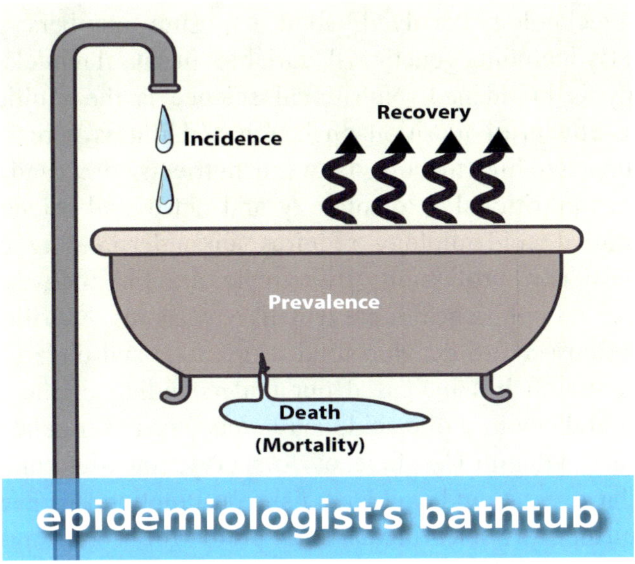

Figure 1.2: *Disease outbreak can be measured by prevalence and incidence levels.*

Source: https://ps-af.facebook.com/publichealthagency/photos/our-latest-blog-discusses-incidence-vs-prevalence-and-the-epidemiologists-bath-tu/3554666981234578/

1.2 HISTORY OF NUTRITIONAL EPIDEMIOLOGY

Before pursuing epidemiology as a primary field of study Dietary exposures have a significant impact on the development of health problems. The assessment of these exposures and the research on the relationship between exposure and outcome is central to nutritional epidemiology. As scientists got a greater grasp of the role minerals and vitamins to play in health and illness, the subject of nutritional epidemiology was transformed in the early twentieth century. The function of exposure in chronic illness became better recognized in the latter half of the twentieth-century utilization of nutritional epidemiology data has enabled tremendous improvements in science and society since then (Wittwer et al., 2011). Epidemiological methodologies were used to study the relationship between nutrition and disease long before there was a scientific agreement.

As the technology evolved, food exposure numbers grew more dependable. By including genetic risk variables in causal models, nutritional epidemiology is broadened. Nutritional science is the multidisciplinary study of the effects of nutrition on health and illness across a person's lifespan. Understanding the link between nutrients, diet, and the human body is where nutritional epidemiology and nutritional science intersect. Grasp nutritional epidemiology requires an understanding of nutrition science. These two professions investigate the link between diet and illness to develop public health preventative measures. Nutritional science research is often used to develop food restrictions and dietary guidelines. Nutritional research has increased our understanding of the relationship between diet and health. Folic acid insufficiency has been related to neural tube anomalies, vitamin C deficiency to scurvy, trans-fat consumption to cardiovascular disease, and excessive fish consumption to a decreased risk of preterm birth. More scientific data and proof will become available over time, increasing the likelihood of successful intervention and prevention. Nutritional epidemiology research provides a solid platform for nutritional discoveries.

Figure 1.3: Nutrition research.

Source: https://thecounter.org/implicit-bias-food-nutrition-studies/

This inquiry into the link between diet and health focuses on chronic disease. They provide a complete picture of how nutrition affects and protects people and groups' health and well-being. Because of measurement errors and unpredictability, there is much dispute on how to consistently and appropriately estimate exposures. To develop medicines and public health policies, nutritional epidemiology study designs must show a clear relationship between nutrition and illness (Van Breda et al., 2015). Observational and experimental research with relevant study designs includes ecological, cross-sectional, cohort, case-control, and clinical trials. Snow is widely regarded as a pioneer in the study of epidemiology as a result of his meticulous work documenting the outbreak of cholera in London in 1854 and pinpointing the cause as the Broad Street pump. Because there was a lack of information about cholera-causing agents, this evidence-based finding was all the more remarkable because it gave an effective technique for preventing the disease.

This case study demonstrates that understanding the molecular mechanisms behind the relationships between exposures and outcomes is not essential when studying causality. Before Snow, a man called Percival Pott observed a high prevalence of scrotal cancer among chimney sweeps, leading him to conclude that their occupation may have exposed them to chemicals that cause the disease. This data set demonstrates how epidemiological concepts may be utilized to identify occupational disorders and begin the process of determining their causation. Richard Doll and Wynder employed well-known epidemiological approaches to establish a relationship between cigarette smoking and bronchogenic cancer in the mid-twentieth century.

According to a thorough study, smokers and nonsmokers have a considerably different risk of acquiring lung cancer, and this difference has been demonstrated to be connected with their lifetime exposure to tobacco smoke. While the effects of smoking are constant across demographic groups (for example, males and women), the tar to which they are exposed includes potent carcinogens that can cause cancer. Although no human studies have been conducted or are planned, it has long been assumed that smoking causes lung cancer. Even in the absence of experimental evidence, such as clinical trials, epidemiology can identify illness causes and preventative strategies (Su & Lu, 2012). Researchers assign exposures in experimental investigations, but exposures are simply observed in observational studies. Because of the risk of exposure in observational research, experimental research can give more evidence for the influence of exposure on outcome. Observational studies, on the other hand, take less time and money to complete.

Over time, observational research may find fresh and unique facts that would be difficult and costly to get through experimental experiments (nutritional disorders develop over time). [However, certain nutritional-disease links pose ethical problems.] Data from several sources influence nutrition policy and practice decisions to guarantee their accuracy, reliability, and validity. The question and the method utilized to perform the research will determine exposure. It can be used by people or organizations from the past or the present, with objective or subjective effects. This word refers to food, including nutrients and no nutrients, as well as the social environment, in nutritional epidemiology research.

Figure 1.4: Nutritional pyramid.

Source: Corsello A, Pugliese D, Gasbarrini A, Armuzzi A. Diet and Nutrients in Gastrointestinal Chronic Diseases. Nutrients. 2020; 12(9):2693. https://doi.org/10.3390/nu12092693

The data can be used to measure the impact of various exposures. In nutritional epidemiology, continuous or discrete variables form the disease state, anthropometric condition, or physiological condition. Evidencing the impact of food and nutrition on disease genesis and prevention is one of the fundamental aims of nutritional epidemiology research. Quality, volume, and balance, as well as those elements that affect food after consumption, must be addressed. Epidemiological research begins with a specified, realistic, and relevant study population and purpose. Secondly, a strategy for measuring exposure and result must be selected and implemented appropriately, followed by a thorough investigation (Singh & Sharma, 2019). The exposure and the intended consequence are analyzed to establish the reliability of the relationship. The findings of well-executed research with a strong basis, comprehensive methodology, and ethical standards will be used to enhance health care. No approach does not demand a previous study of the literature. A variety of nutritional epidemiology study methods have their advantages and disadvantages depending on the situation at hand.

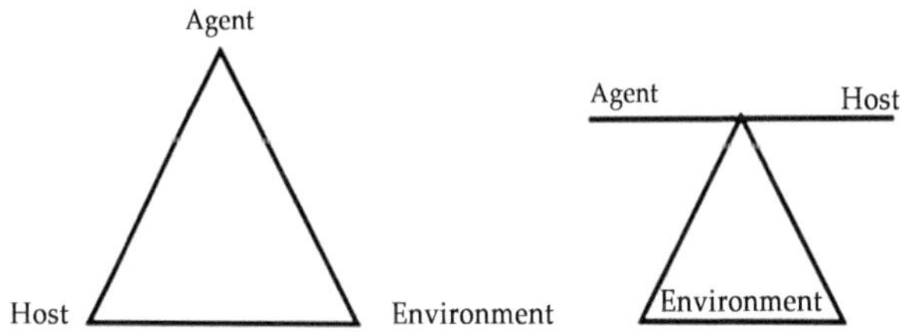

Figure 1.5: Epidemiology concept.

Source: https://www.cdc.gov/csels/dsepd/ss1978/lesson1/section8.html

Environmental investigation and ecological research examine how risk-modifying variables impact the health outcomes of a group of people in geographical or temporal ecological environments. There is a danger that ecological studies may fail to capture the true interplay between individuals in large groupings. Even though ecological studies employ geographical information to examine the spatial context of illness and exposure, systemic heterogeneity in categorization language is possible. Cross-sectional study. Researchers examine how people are exposed and what occurs as they go about their regular lives in a cross-sectional study. When researching the relationship between disease and diet, cross-sectional studies give a snapshot of sickness prevalence in a population at a certain point in time.

Because of its capacity to examine a wide variety of outcomes and exposures, cross-sectional studies can aid in health resource planning and allocation by examining the sickness burden in a given population. When determining the success or failure of a project, there are several aspects to consider. Respondent prejudice is a primary factor in nonresponse, which leads to skewed data (Sikalidis, 2019). Case-control studies (in which outcomes are measured now and past exposure is recorded) (i.e. measures outcomes now and documented prior exposure) Case-control studies are individual observational studies that focus on the outcomes (i.e. Measures outcomes in the present and past exposure is established). Either group, referred to as controls or cases, receive either exposed or unexposed therapy (disease). Case-control studies can be used to research uncommon diseases over time, but they are confined to analyzing a specific outcome and are vulnerable to bias if the control groups chosen are not representative of the general population, leading to incorrect findings.

Cohort studies: A cohort study is a sort of observational-individual research that assesses current exposure over time and then predicts future consequences. A cohort study can investigate a wide variety of outcomes for a single exposure. It is critical in nutritional epidemiology research to be able to quantify both the incidence and prevalence of outcomes following exposure. Cohort studies, on the other hand, are costly and time-consuming to conduct. As a result, mistakes in data collection or confounding variables cannot be corrected before the results are determined (Riscuta, 2016). There are two sorts of trials: clinical trials and community trials. Experimental research involves direct interaction with people or populations (clinical community). Clinical trials usually contain tests and procedures on individuals assigned to various treatment groups. Clinical trials can be conducted to evaluate new cures, medications, and procedures.

Clinical studies, on the other hand, expose participants to the risk of adverse effects and accidental damage from the intervention and should be considered only when the supporting data is good enough to warrant it. In community trials, people with and without illnesses are randomly assigned to different treatments. It does not, however, account for individual variances. The consequences for the community nutritional epidemiology has historically had a significant impact on society, health, and the economy. Nutritional epidemiology research is used to provide dietary advice for illness and cancer prevention.

Figure 1.6: Epidemiology and population health.

Source: https://www.publichealth.med.miami.edu/divisions/epidemiology-and-population-health-sciences/index.html

They have made substantial contributions to nutrition and health policy as a result of their dependence on well-researched data. Because of the observational data, food fortification and restrictions on specific dietary items were made possible. Since the 1970s, people's health and well-being have improved as a result of these changes, via both prevention and improvement. Nutritional assistance has been demonstrated in studies to enhance cancer patients' quality of life by minimizing side effects, boosting treatment response, and decreasing the chance of cancer recurrence. As more individuals are diagnosed with a wide range of diseases and ailments, the healthcare system is becoming overburdened, making it increasingly difficult to operate at full capacity. Nutritional epidemiology provides the scientific foundation for public health nutrition (Riscuta & Dumitrescu, 2012).

The purpose of nutritional epidemiology is to provide information on how to address vitamin imbalances that cause illnesses including anemia, goiter wasting, and stunting. Because exposures have particular features, nutrition and sickness go hand in hand. The necessity of correlating exposure with specific causal pathway stages is highlighted in public health recommendations. To establish a definitive relationship between nutrition and sickness, evidence from several research types is necessary. Cell culture tests, animal studies, and human metabolic investigations all give mechanistic data suggesting the role of a certain dietary component in carcinogenesis. However, no one dietary item has been demonstrated to be a cause or a treatment for cancer in humans. Human research, ideally randomized controlled trials, is the only one that can produce proof. However, such testing is not always feasible. Long-term research is difficult to support since cancer takes so long to grow. It is likewise morally unacceptable to investigate the link between a potentially dangerous exposure and cancer in humans. For these reasons, observational nutritional epidemiology research can suggest a diet-cancer link.

This chapter provides an overview of nutritional epidemiology concerning cancer, with a focus on research design, dietary assessment/analysis, and data interpretation. The bulk of dietary variables must be potent and consistent to be categorized as important carcinogens or anticarcinogens. Some local practices, such as feeding Chinese-style salted fish to newborns, have been related to the development of nasopharyngeal cancer, as have the use of alcohol and the intake of aflatoxin-contaminated foods (such as feeding Chinese-style salted fish to infants, which causes nasopharyngeal cancer). Some culinary characteristics have been regularly connected to cancer rates during the last two decades, while opinions on the validity of the research remain divided (Ray, 2014).

In the 1980s, foods abundant in fiber, fruits and vegetables, vitamins A, C, and E, and beta-carotene were shown to be protective against heart disease, whereas excessive fat, animal protein, and calorie consumption were found to be risk factors. Recent research has identified red and processed meat as risk factors, whereas folate and other B vitamins, vitamin D and calcium, selenium, and some non-nutritional components of fruits, vegetables, and whole grains have been discovered as protective factors (e.g., phytoestrogens and isothiocyanates found in cruciferous vegetables).

Figure 1.7: Laboratory tests by a female scientist.

Source: https://www.healthcare-management-degree.net/faq/what-is-epidemiology-and-what-does-an-epidemiologist-do/

Fats appear to be less harmful to health than previously assumed, and more specific substances and processes are being studied. Both animal and vegetable fats contain saturated fat, monounsaturated fat, trans-fat, and other types of fat (e.g., saturated fat). Some people may benefit from increasing their intake of omega-3 fatty acids (found in fatty fish). The American Institute for Cancer Research and the World Cancer Research Fund examined diet and cancer research. Some nutritional supplements may have different effects than foods that naturally contain them, and some may even be harmful to your health (Qadeer & Ashhar, 2014). Although beta-carotene has been related to a decreased risk of cancer in observational studies, a 12-year randomized trial revealed no benefit, and persons who got b-carotene supplements were shown to have a greater risk of getting lung cancer in two shorter trials. Aspirin may lower the risk of colorectal cancer, albeit this may take a decade or more to become apparent. According to the majority of research, dietary adjustments might save one-third of cancer deaths. Only obesity-related malignancies are undeniably preventable with a good diet. Even if everyone agreed on what meals are necessary, a dramatic alteration in national eating patterns would be difficult. Obesity has long been linked to a higher risk of cancer. Even while postmenopausal breast cancer, endometriosis, gallbladder cancer, and kidney cancer have the most evidence, other sites contribute to total cancer risk. The biggest prospective cohort of nonsmokers in the United States, where obesity is especially prevalent, provides the most compelling evidence of a relationship between BMI (weight divided by height squared) and cancer mortality. If the observed higher mortality is a direct result, their estimates show that around 10% of all cancer deaths among nonsmokers in the United States might be avoided if no one had a BMI of more than 25 kg m2 (7 percent in men and 12 percent in women) It is unknown, however, how much a person's risk can be decreased by lowering weight if they are currently obese. People who are either too thin or too overweight are more likely to die from noncancerous conditions. As obesity grows more frequently in Western countries, the frequency of weight-related malignancies is predicted to climb in the future years.

Figure 1.8: Dietitians play a crucial role in healthcare delivery.

Source: https://dutable.com/2019/02/20/the-role-of-dietitians-in-health-care-delivery/

1.3 OBSERVATIONAL STUDIES

Unfortunately, the discipline of nutritional epidemiology is hampered in its efforts to establish solid correlations between food consumption and sickness risk. As a result, these obstacles are approximately listed in order of importance: If present or recent dietary consumption differs from intake during the critical era for illness development, we may be unable to find food-disease links. A population's nutritional consumption may be quite stable. In postmenopausal women, a suitable disease risk range for establishing a relationship between breast cancer and dietary fat consumption is 25–40% of total caloric intake. A comparison may be made between the research on whether smoking causes cancer and the study of men who smoke 1.5 packs per day against those who smoke one pack per day (Pauletto et al., 2020). It is particularly difficult to separate the effects of one molecule from those of other dietary components since the diet consists of numerous foods and nutrients, all of which have a direct effect on the body. Diet and other risk factors including hormone status, obesity, and hypertension are tightly linked. Because of these connections, it is difficult to control for confounding circumstances (some of which may be causative).

Although it is obvious that many types of random and systematic error impede our capacity to discover correlations between food consumption and sickness risk, the majority of the measurement aspects of contemporary dietary self-report instruments remain unknown. The laboratory procedures employed are governed by the sort of study being undertaken. The frequency of nutritional epidemiology field tests in developing nations will be restricted by sample size, the suspected incidence of deficiencies, local laboratory conditions, the availability of competent employees, and budgetary constraints. Some of the most regularly evaluated markers include blood hemoglobin, hematocrit, serum iron, TIBC, serum ferritin, blood protoporphyrin, serum albumin, plasma transport proteins, and serum zinc. Other vitamins in the blood or urine, as well as certain hormones (thyroxine and TSH), and minerals, may create problems (urinary iodine). When selecting laboratory tests for population research in industrialized nations with strong laboratory facilities, there are several considerations to consider. Without additional factors for evaluating protein status, plasma transport proteins would be the only variables to be considered.

Figure 1.9: Epidemiology data illustration.

Source: https://onlinedegrees.unr.edu/blog/epidemiology-vs-biostatistics/

Lipid pattern studies are performed to measure EFA status; the test utilized depends on whether recent consumption or long-term status is more essential. Serum cholesterol, triacylglycero lipids, and maybe lipoprotein fractions are also measured. Because preclinical deficiencies are widespread in industrialized settings, a battery of tests should be undertaken in the absence of dietary information. The majority of the recommended and best available approaches were employed in the Third Nutritional Health and Nutrition Examination Survey conducted in the United States (Peña-Romero et al., 2018). This is also true for the current food and nutrition surveys on particular demographic groups done by the United Kingdom government. Dietary epidemiology studies how one's food influences one's health. This comprises investigating the direct and indirect relationships between various dietary components, nutrients, and pollutants and various health outcomes or sickness symptoms.

Exponent researchers use their expertise in nutritional epidemiology, as well as their knowledge of national health databases, applied statistics, and risk assessment, to investigate the impact of these factors on health outcomes such as nutritional adequacy, diet quality, disease incidence, and biomarker levels. To study these links, observational data studies, including meta-analyses, and the preparation and execution of substantial, evidence-based literature reviews might all be employed. The food and beverage industry, its associated trade and commodity groups, and food lawyers can use nutritional epidemiology to support regulatory submissions to government agencies such as the US Food and Drug Administration (FDA), provide safety evaluations of food ingredients, respond to litigation, and develop persuasive public health messaging about their products. Traditional study on diet and sickness has concentrated on a few essential foods or minerals rather than the entire diet. This level of concentration is essential in the food and beverage business.

Proactive safety assessments of food items or ingredients, as well as scientific research into the effects of certain dietary components on health, are two examples of these activities. Nutritional epidemiology may also be used in response to litigation and the need to defend a specific food product against adverse charges. Toxins have been discovered in several food and beverage items that the food and beverage industry has justified citing nutritional epidemiology, according to the Office of Environmental Health Hazard Assessment (OEHHA). Using nutritional epidemiology data, public health professionals may establish if a food contaminant is associated with poor health outcomes and, if so, whether the negative consequences exceed the advantages of the product's overall function in a healthy diet (Banerjee et al., 2015). The relevance of this study creates new opportunities for corporations to use nutritional epidemiology to examine how a product fits into a consumer's eating habits and the overall impact on health. Every five years, the Departments of Health and Human Services and Agriculture create the Dietary Guidelines for Americans (DGA).

With the publication of the 2015-2020 Eating Guidelines for Americans, the emphasis switched from analyzing foods and minerals on their own to consider a person's whole eating pattern. The 2020-2025 version of the Dietary Guidelines for Americans (DGA) will place a greater focus on nutrition, looking at what defines a "healthy" diet and the possible advantages of adopting it for the US population. Food and beverage companies, as well as trade and commodity associations, have an opportunity to examine how their products fit into healthy dietary patterns and how they contribute to Americans achieving nutrient sufficiency, improving overall diet quality, and lowering the incidence of chronic disease. According to a recent Forbes3 article, a team from Exponent undertook thorough research to assess how much money might be saved in US healthcare by lowering the number of chronic diseases in the population. The findings of this National Dairy Council-funded study were reported in the Journal of the Academy of Nutrition and Dietetics4. According to a recent study from Exponent and the American Beverage Association, those who consume low- and no-calorie sweetened beverages, as well as zero-calorie unsweetened liquids, do better than those who drink sugar-sweetened beverages.

To address this issue, NHANES (National Health and Nutrition Examination Surveys) provide cross-sectional data on food intake and composition of the current US food supply. Our studies also necessitated a professional understanding of nutrition, diet, and health concerns comparable to that of the FDA, USDA, and Environmental Protection Agency (EPA) in monitoring and enforcing food safety and making policy decisions. We published our findings in the journal Nutrients, which is open to the public. They demonstrate the Exponent's competence in nutritional epidemiology and have now been published in the scientific literature to support the 2020-2025 DGA improvements in this area. Nutritional epidemiology has always had its skeptics, but there are now many more as a result of recent breakthroughs in the subject (Piroddi et al., 2017).

According to Ioannidis, "the ability of US nutritional monitoring to foresee population trends in caloric intake and establish empirically informed public policy about diet-health links is quite confined." Several dubious meta-analyses have also promoted contradictory nutritional and health messages, such as the notion that being overweight reduces the risk of mortality from any cause and that replacing saturated fat with polyunsaturated fat has no effect on cardiovascular risk. Findings like this are not only intriguing but also potentially detrimental, because they may be misunderstood as signals or transmitted to the broader public via popular media, leading to the majority of the population adopting potentially hazardous habits. Following the publication of the second meta-analysis. In actuality, randomized controlled trials are not a panacea for nutrition and chronic illness research, and their conclusions may be deceptive. Food consumption is a complex exposure of interest in nutritional epidemiology, with interactions and synergies between various dietary components that may be difficult to evaluate using a linear drug trial technique. Nutrition poses a distinct set of obstacles due to the difficulties and impossibilities connected with pharmacological studies involving food treatments, such as achieving high compliance and blinding.

As a result, nutritional epidemiology poses unique design and analytic obstacles, and a grasp of nutritional science and its methodological foundations is essential to appreciate the intricacies of nutritional epidemiologic research. The purpose of this book is to clarify myths about nutritional epidemiology, highlight the obstacles that the profession faces, and evaluate the role of nutritional research in determining public health policy. Research addresses five critical difficulties in nutritional epidemiology research. Is it feasible to accurately assess individuals' and communities' nutritional intake? (Pal, 2018). Individual meals contain numerous and sometimes poorly defined components that are ingested in varying amounts and combinations by different people, making nutrition evaluation challenging in free-living communities. Nutritional impacts are seldom one-to-one; in most, but not all, circumstances, everyone is "exposed" to some extent.

Figure 1.10: Mental health foods.

Source: https://www.eatright.org/health/wellness/healthy-aging/4-types-of-foods-to-support-memory

Nutrition is a variable exposure when it comes to how individuals eat and how their diets vary over time. When employing dietary assessment methodologies, some degree of inaccuracy is to be expected, since this can be produced by random fluctuations in a person's diet from day to day, day to day, and season to season, or by purposeful activity, such as deletion of items when collecting data. Several methods for estimating the amount of food consumed by free-living societies have been proposed. However, when numerous characteristics are included, these techniques have demonstrated great validity (Neeha & Kinth, 2013). Even though each approach has its own set of limitations, it may be employed in certain instances due to its unique properties. Multiple-week diet records are considered the gold standard for giving nutritional information since, unlike other approaches, they do not rely on recollection.

Diet records have been unable to be used in large-scale epidemiological studies due to the high participant burden and cost of maintaining them, but their ability to accurately ascertain detailed dietary information makes them useful in validation studies of other diet assessment methods and monitoring trial compliance. Another disadvantage of maintaining a food diary is that it may lead to dietary changes that impair the accuracy of the data. Nonetheless, food diary intakes have been demonstrated to be very similar to those from other 24-hour recalls. A competent interviewer must recollect all meals consumed in the previous 24 hours or calendar day in person or over the phone. Even when participant memory increases the likelihood of measurement error, a competent interviewer can offer nutritional data that is as reliable as a diet record.

This strategy has been utilized often in nutritional intervention research. It tracks nutrient consumption patterns as part of national surveys. Nutritional evaluation methods that rely on food composition tables are not without danger. The nutritional content of a product can be impacted by the season, where it was produced and collected, as well as the storage conditions, processing, and cooking procedures used. This may be irritating to varying degrees depending on the vitamin. Assuming that these oscillations have no substantial influence on estimated intakes for certain elements, such as dietary fat, the variance might result in calculated intakes that range significantly from actual intakes for others, such as selenium. However, in general, this type of error does not significantly impede the ability to rank individuals based on nutrient consumption to explore the association between health outcomes and nutrient consumption (McNamara, 2010). Given this, we must continue to enhance the quality of food composition databases, as calculating nutritional content from consumption figures is challenging. Researchers can utilize biomarkers to identify how much biological material a person has contributed. Biomarkers include doubly labeled water, urine nitrogen, and 24-hour urinary salt and potassium levels. Biomarkers include serum and plasma folate, as well as trace minerals in toes. Because biomarkers provide an objective evaluation of consumption, there is no need to rely on self-reporting.

Biomarkers, on the other hand, have been impeded in their widespread usage due to flaws. Because of a lack of biomarkers for many foods and nutrients, as well as a lack of specificity in their evaluation, biomarkers are difficult to employ as trustworthy predictors of a person's long-term intake. Biomarkers have mostly been employed in nested case-control studies and short trials to assess the relationship between nutrition and disease. A variety of less-priced, self-reported diet evaluations that may be evaluated utilizing biomarkers are available. Food Frequency Questionnaires (FFQs) are made up of a food list and a frequency answer section in which participants indicate their frequent consumption of each item over a specific period, usually a year. This technique is the most widely used in large observational studies to estimate intake because of its low burden on participants and ability to capture typical long-term food consumption. Recurrent examinations of these features are required for demonstrating long-term changes in diets.

Nutritional epidemiologists have uncovered and documented food consumption data measurement mistakes during the last 50 years. Because of the large diversity in food consumption between people on any given day, a single recall like the one utilized by Archer et al. might lead to incorrect findings. They are more likely to record extremes in food consumption in their research than the typical person's current intake, which means that any single person's recall might be ridiculously high or low. As a result of this random fluctuation, variance is overestimated and the distribution flattens, resulting in an increase of people at the extremities of the distribution. To limit the chance of in-person mistakes, several 24-hour recalls on nonconsecutive days are advised. To address some of these difficulties, a second 24-hour recall was included in the NHANES beginning in 2002, and additional epidemiologic studies that employ 24-hour recollections to evaluate diet now include multiple duplicate evaluations for each participant (Mustapa et al., 2020). Diet assessment errors, on the other hand, do not have to be wholly random.

Figure 1.11: Epidemiology graphic design.

Source: https://theconversation.com/how-to-model-a-pandemic-134187

Over- or under-reporting due to social approval or desirability (acting in specific ways to earn social praise) is a systematic source of variance (reacting in certain ways to avoid social criticism) As a result of these factors, Archer et al. discovered underreported energy consumption. Nutritionists have long acknowledged the underestimating of calorie consumption from self-reported data, and several advances in methodology have been explored to eliminate this source of measurement error. If an omission or estimating mistakes in portion size are the reasons for under-reporting, the study's results will be the same. Furthermore, breakthroughs in 24-hour recall approaches that minimize these errors have been established, such as the USDA 5-pass strategy, which aims to reduce food omission and assist participants in reporting proper portion amounts by employing visual cues.

A statistical model that considers total energy intake might also be utilized. In analytic epidemiology, we are interested in how dietary composition impacts disease risk rather than how much energy/macronutrients an individual consumes since this is the most adjustable by individuals or communities. Reducing the quantity of energy consumed is a common strategy in nutritional epidemiology. Systematic sources of under-or over-reporting can be minimized to a lesser extent by controlling for energy intake. Measurement error affects a wide range of behavioral exposures and biomarkers, including physical activity, which Archer and his colleagues are well-versed in due to their extensive work in the field of physical activity epidemiology (Masotti et al., 2010). Although every approach has faults, there is substantial evidence that dietary measures in national surveys are trustworthy and accurate. Dietary data obtained in this manner cannot be utilized to guide public policy and is deceptive. NHANES statistics provide for a minor portion of the overall data used to define national standards and public policy.

The primary goal of analyzing national survey data is to determine how much food the average individual consumes and how that trend is evolving. Using NHANES data, a representative sample of 29,124 persons aged 20–85 years old was recently evaluated, and the results revealed trends in dietary quality in the US adult population from 1999 to 2010. According to the Alternate Healthy Eating Index, the higher socioeconomic class was associated with better nutritional quality, and the difference between affluent and poor expanded with time. According to the findings, nutritional strategies should be enacted to increase food quality and eliminate health inequities. Because of their low cost and minimal participant load, self-administered, computer-processed food intake surveys are the only choice in most large cohort studies.

As a result, when used to estimate average normal consumption throughout the prior year, FFQs show lower random within-person variance than other nutritional assessment methodologies. As a result, there is a better knowledge of long-term food intake, which is important in the majority of diseases. Because they are dependent on memory, FFQs may have a larger degree of measurement error over time than memories and records. A few days of diet records or recalls may be sufficient for several nutrients, provided the days are spaced out across the whole FFQ reference period (Mathers, 2017). "For certain nutrients, a few days' worth of diet records or recalls (spaced out across the whole FFQ reference period) may be sufficient." When compared to a reference, FFQs have been demonstrated to have adequate validity, with correlation values ranging from 0.4 to 0.7 for specific nutrients or meals.

The use of repeated FFQs in long-term prospective cohort studies, as well as controlling for total calorie consumption, improves the validity coefficients even further. Where biomarkers are available, triangulation processes may produce higher estimates of correlations between FFQ consumption and actual intake than the most often used reference methodology, which is keeping long-term food records. These validity factors may be used to account for measurement error in epidemiological analysis, and their use in more advanced analyses is expanding. Researchers were able to derive accurate results from large cohort studies using these methodologies, which made use of FFQ data. Food intake data measurement errors have been greatly decreased, but additional progress in dietary assessment and measurement error correction is required to advance the field forward. While progress has been achieved in recent decades, nutritional epidemiologists can now securely gather and use dietary information for both individuals and groups due to the use of repeated diet evaluations across time.

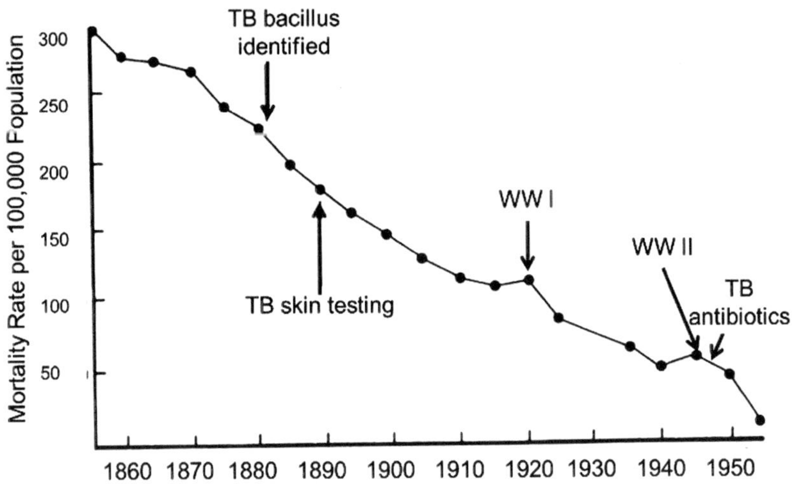

Figure 1.12: Epidemiology data illustrations of tuberculosis.

Source: https://sphweb.bumc.bu.edu/otlt/mph-modules/ep/ep713_descriptiveepi/ep713_descriptiveepi_print.html

Do nutritional epidemiology studies help in determining illness etiology? Nutritional epidemiology, opponents argue, relies too largely on observational data, which is deemed less credible than experimental evidence in identifying causality. Despite their position at the top, randomized trials with unequivocal findings are seldom the best and most practical option to settle nutritional epidemiology problems concerning the long-term effects of certain meals or nutrients (unless they can be packaged in a pill). Exponent's toxicologists and epidemiologists are recognized as specialists in food safety and nutrition in the United States and across the world (Loor et al., 2015). Performing nutritional epidemiology analyses and systematic reviews has assisted clients with regulatory filings to government agencies, product safety investigations, legal actions, and the production of an effective public health message surrounding their goods. It is feasible to show causation if several exposures (e.g., dietary patterns, food intake, and biomarkers) are studied.

A Mediterranean diet, for example, has been demonstrated to be heart-healthy in both observational and randomized studies. Fish, almonds, olive oil, fruits, and vegetables are all high in omega-3 fatty acids, which reduce the risk of heart disease. This conclusion is supported by biomarker studies and analysis of important dietary components such as polyphenols and unsaturated fats. According to this research, which has been replicated several times, a Mediterranean diet provides actual benefits in terms of lowering the risk of cardiovascular disease. In nutrition and health research, prospective cohort studies are not the only technique to show causality. To make a therapy or intervention choice, animal research, human mechanistic investigations, and studies of hard endpoints, as well as randomized trials of intermediate outcomes, are employed. If the results of these tests are consistent, causation can be established. When trans-fats are taken, a rise in LDL cholesterol and a decrease in HDL cholesterol have been demonstrated to be detrimental to intermediate endpoints such as blood pressure (Ladeira et al., 2018). In observational research, an increase in trans-fat consumption was connected to an increased risk of ischemic heart disease (taken together, these data construct a solid case for the negative effects of trans-fat consumption on heart disease). This technique focuses on contextualizing individual research rather than examining them in isolation.

Earlycohortstudies,ratherthanrandomizedtrialswithobjectiveoutcomes, can be used to demonstrate causation with high certainty. Confounding and other types of bias in research studies can be reduced by using a high-quality study design, extensive statistical analysis and interpretation, and frequent replications of the findings. In the absence of large randomized controlled trials with specified goals, Hill criteria may be used to establish causation, and data from other study kinds and demographics can back this up. Do researchers investigate diet-disease links in the same manner that they do medication interactions? According to some academics, retrospective causal inference (RCT) is the apex of causal inference. Because of the occasional blinding of dietary treatments in the pharmaceutical industry, the impact of an intervention may be attributed to treatment assignment information rather than the nutritional component of the intervention. Dropout rates are higher in nutritional intervention studies, especially when the intervention is protracted or stringent, as in pharmacological trials.

In the case of random loss to follow-up, a high dropout rate reduces the power of the study. When the dropout rate varies according to treatment and outcomes, there may be systematic bias. The nutritional study is hampered by participant noncompliance or failure to adhere to their recommended treatment plan. The longer the trial lasts, the more probable the defendant will disobey. After eight years of follow-up in the Women's Health Initiative (WHI), the majority of individuals randomly assigned to the low-fat group failed to meet their 20 percent fat reduction objective. In terms of plasma concentrations, HDL cholesterol and triglyceride levels did not differ between the low-fat and control groups during the trial. As a result, the WHI was unable to demonstrate that a low-fat diet had a causal effect, and its null results were frequently ambiguous (Kussmann & Affolter, 2009). This complicates and restricts the significance of randomized studies of nutritional treatments, which are not affected by post-baseline confounding and selection bias. It's also feasible that long-term research will become outdated. It is widely assumed that reducing fat consumption will reduce the risk of breast cancer and cardiovascular disease. In the early 1990s, the Women's Health Initiative (WHI) was formed to do a study on this topic. The basic premise of the experiment was called into question by unanticipated clinical and epidemiological outcomes, notably in the field of cardiovascular disease (CVD).

CHAPTER 2

RELATIONSHIP BETWEEN HUMAN GENOME, NUTRITION AND HEALTH

CONTENTS

2.1 INTRODUCTION

It has been common knowledge for a very long time that the food that one consumes may have an impact on their overall health. The ancient Greek physician Hippocrates, who lived around 400 B.C., is credited with giving the following piece of advice to his contemporaries in the medical field: "Leave your prescriptions at the pharmacy if you can treat your patient with food." In addition, it is common knowledge that various persons have varying dietary requirements (Kussmann & Van Bladeren, 2011). This is a well-established fact.

A person's food can have an effect, either directly or indirectly, on the beginning of the course of a disease in that individual. The interaction of foods and meals with genes has an effect that is favorable in the vast majority of situations; nonetheless, this interaction might on occasion have a fatal consequence.

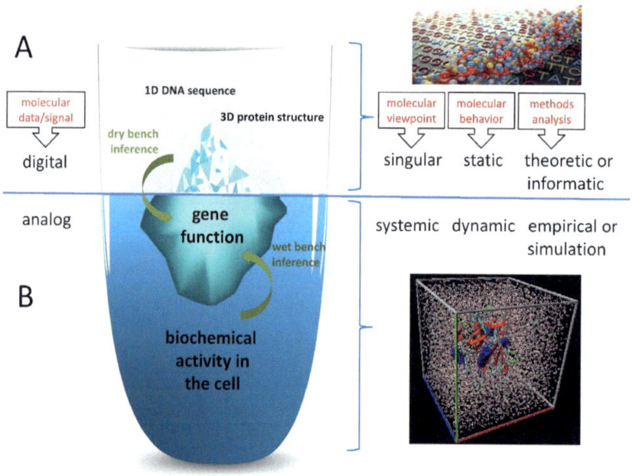

Figure 2.1: Illustrative model of the human genome.

Source: https://www.researchgate.net/figure/Epistemological-and-methodological-comparison-of-the-perceptions-biases-technical_fig1_289122369

Both environmental and genetic variables may have an impact on human beings, and it is essential to keep both sets of influences in mind if one wants to maintain optimal physical health. The overwhelming bulk of the research that has been conducted in the past has focused either on the part that genes play or on the influence that environmental circumstances have, but rarely on both at the same time. Recent research has investigated the relationship that exists between a person's genes and the diet that they consume.

To achieve optimum health, the academic discipline of nutrition science explores how nutrients influence the homeostasis of the body's cells, tissues, and organs. To be successful, this research must comprehend not only the genetic level but also the molecular level, the level of protein synthesis, and the level of the metabolic profile of nutrient-dependent interactions. As a direct result of this, the field of nutrition research has expanded to encompass, in addition to epidemiology and physiology, molecular biology, genetics, and nutritional genomics (Kang, 2012). Previously, the scope of nutrition study was limited to just those two areas.

The study of how the foods we consume and the genes we inherit interact with one another is referred to as "nutrigenomics," and it is a topic of academic research. People who come from a wide diversity of genetic backgrounds are the subject of a substantial amount of research and attention that focuses on how varied diets affect their health. The purpose of this method is to provide individuals with the kind of nourishment that is most suited to meet the requirements that are unique to them. In the not too distant future, scientists, legislators, and health industry professionals are working together to develop personalized nutrition guidance and the manufacturing of functional foods.

These foods have the potential to improve an individual's health depending on the requirements that are unique to that individual. For example, the meals that we eat are linked to the foods that our ancestors ate, and they also have an impact on the growth and development of the cells that make up our bodies. In addition, it is akin to how the food we eat influences our physical growth in a manner that is equivalent to how our genetic composition affects us. With the use of nutritional profiles, researchers will be able to discover the factors that explain individual differences in dietary needs as well as how effectively individuals react to food-related therapies. Nutritional genomics refers to the study of how a person's genetic makeup influences how their body reacts to the foods that they eat. This field of research has been increasingly popular in recent years (Irimie et al., 2019). People are considered to be distinct variants of the same gene, even though they all come from the same gene pool. Some people's responses to specific meals, for instance, are more adverse than others' due to variances in their genetic make-up. This is because some people have a more sensitive digestive system than others. The consumption of particular food concoctions can be a contributing factor in the development of a wide variety of illnesses and conditions. Genes are the components of an individual's genetic makeup that are handed down from one generation to the next.

Genes are inherited by offspring from their parents. Not only are our genes responsible for the generation of proteins within our bodies, but they are also responsible for the particular qualities that differentiate us from one another. The previous list demonstrates that there are overt and covert parts of a person's biology that can impact their health. For example, the color of a person's eyes and the shape of their nose are both examples of overt features of a person's biology that have this capacity. It is estimated that the human genome has anywhere between 20,000 and 25,000 genes. In the year 1990, scientists made the initial steps in locating and sequencing the genes that are contained in the human genome (Hurlimann et al., 2014). In the year 2000, researchers generated the first-ever complete map of the human genetic makeup. This map was made possible by advances in DNA sequencing technology.

In the field of science known as Nutrigenetics, the subject of how our genetic make-up impacts our reactions to various meals, beverages, and dietary supplements is examined. Nutrigenetics is one of the newest scientific disciplines. In the course of this research, an individual's distinctive genetic make-up was dissected to establish the role that it plays in determining how a person reacts to various kinds of food. It has been known for a very long time that specific diets may have different impacts on different people and that some people are more sensitive to the effects of certain diets than others. If you suffer from lactose intolerance, you run the risk of suffering stomach pain after ingesting certain dairy products; however, this is not a problem for everyone. Lactose intolerance is the condition in which a person is unable to digest the natural sugars that are present in foods made from milk.

Lactose intolerance is differentiated from other digestive illnesses by the fact that it is caused by a deficiency in the activity of the lactase gene. This might cause pain in the stomach, including bloating, gas, diarrhea, and sensations that are comparable to nausea. Individuals who have phenylketonuria do not generate the enzyme phenylalanine hydroxylase, which makes it difficult for these people to metabolize the amino acid. PA will almost invariably be converted to tyrosine in the body of a person who is otherwise in good health. On the other hand, those who have PKU have an increased production of phenyl pyruvic acid within their bodies. If there is a high concentration of this acid inside the body, there is a chance of complications relating to brain function, including mental retardation and convulsions (Hesketh, 2013).

This risk is increased if the acid is present for an extended period. The end objective of nutrigenetics is to be able to provide people with nutritional advice that is personalized to their unique genetic make-up. This is the ultimate goal of nutrigenetics. Nutrigenetics When discussing the genetics of nutrition, there are a few things that stand out as particularly significant to keep in mind for this topic. To begin , it is essential to admit that the diets of certain people could significantly increase the possibility that those people would have particular diseases. This is a crucial step before we can move on to anything else. A person's regular diet can also have an impact on the expression of genes as well as the structure of their genome.

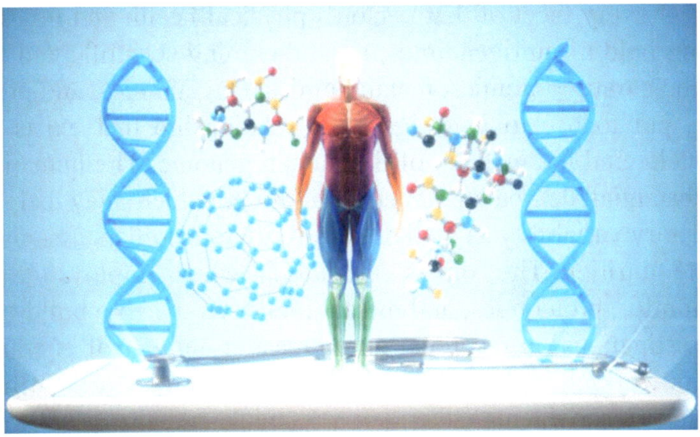

Figure 2.2: Genetic screening.

Source: https://media.istockphoto.com/photos/healthcare-technology-concepts-picture-id1129302149?k=20&m=1129302149&s=612x612&w=0&h=gCdPJ0jwDaRi3sV6qE3F93F_byT8_R8go5RdSBpT9Aw=

There are connections between the expression of genes and the foods that people eat.

Nutrigenomics: The pathways that are depicted in the figure are the ones that foods may use to exert their influence on our genes and alter the way our bodies display our genes genetically, also known as our phenotype. The results of method A demonstrated that nutrients perform the same role as ligands for transcription factor receptors and have an effect on the proliferation of cells when they are in their most healthy state. As can be seen in Route B, a person's metabolism is not the only thing that may be influenced by the food that they consume. Changes in the concentrations of substrates or intermediates can bring about these changes via either the primary or secondary routes, depending on which pathway is being used. Because of this, a person's eating habits may have an influence not only on their genetic make-up but also on the balance between healthy and unhealthy conditions in their lives (Gonzalez et al., 2015). Various genes are considered to be involved in the genesis, progression, and severity of chronic disorders. These genes are influenced by diet. As a result, one's eating habits can play a role in the prevention, treatment, and possibly even cure of chronic illnesses.

The difference between these two names is not always clear when they are being used in a sentence. Like pieces in a jigsaw puzzle, an individual's genes affect every facet of that person's physical health and fitness. In the burgeoning field of nutrigenomics, up-and-coming scientific and technical fields such as transcriptomics, metabolomics, epigenomics, and proteomics are being put to use to investigate the relationship that exists between dietary intake and the makeup of the human genome. The human genome has a substantial degree of genetic variation, which may help explain human dietary variability as well as sensitivity to disorders associated with inadequate nutrition. The "omics" revolution, which employs the genome, transcriptomics, proteomics, and metabolomics as the core building blocks of its investigations, is receiving an increasing amount of attention from researchers working in the field of nutrition. When the Greek suffix "ome" is added to extensive nouns like genes, proteins, and a wide range of other nouns, the idea of "completeness" is communicated to the listener (Guéant et al., 2014). Because of this, genomics goes thoroughly into the research of DNA structure and function, whereas transcriptomics focuses on the analysis of patterns of gene expression. As a consequence of this, genomics differs significantly from transcriptomics in its approach. In contrast, metabolomics explores the disassembly of metabolites into their parts to study the structure and function of proteins, whereas proteomics focuses on the creation and organization of proteins.

There are connections between the expression of genes and the foods that people eat.

Consumption of food can have an impact on the pathways that are involved in signal transduction, which in turn can have an impact on how cells divide and multiply.

The researchers discovered a connection between the activity of cells and the expression of genes inside those cells as a reaction to a broad collection of environmental variables. This relationship was observed in response to the environmental factors. In addition to this, they looked at the role that nutrition plays in the process of cellular differentiation and development. To supplement the research that has already been done on metabolomes and proteomes, it is becoming increasingly important to conduct analyses of the genome or transcriptome (Fenech et al., 2011). The nutritional benefits of nutrigenomic food studies have been given a new perspective as a result of research into the molecular impacts of food components such as carbohydrates, proteins, lipids, carotenoids, vitamins, minerals, and flavonoids. This research was carried out to answer the question, "How do food components affect our bodies?" There are a lot of various applications for nutrigenomics, some of which include research on genetically modified species, the development of an individual's personalized diet, and the defense of the genuineness of food and meals.

Figure 2.3: Genetic screening in detecting diseases.

Source: https://www.theindianwire.com/wp-content/uploads/2021/10/iStock_genetic-engineering-3x2-1.jpg

The genetic make-up of human beings and mice of the Agouti strain is very close to one another. Differences in the expression of the agouti gene are what cause the coat color to look differently at different times. Altering the nutrition of the mother before, during, and after pregnancy is one way to manage these variances and prevent them from occurring. To turn on the agouti genes in newborn agouti mice, food molecules that are linked to agouti genes can operate as a chemical switch if they are in the right place at the right time. This is a possibility because the genes are connected to the molecules of food (Ferguson et al., 2010). Even though the genetic material had been preserved, there had been no sign of activity in the genes for a considerable amount of time. Mice are capable of producing offspring that are born with a brown coat, and these newborns mature into robust and healthy adults. They have enough amount of food.

DNA methylation is a process that occurs in diseases like cancer, and it has been demonstrated to have a substantial influence on the expression of genetic material. According to the findings of Pufulete and team, taking folic acid helps to correct DNA hypomethylation. They have collated their results, and those findings point in this direction. One of the most fascinating and potentially fruitful recent developments in cancer treatment is the investigation of apoptosis delays or delays. It has been demonstrated that flavonoids possess a significant amount of potential as cell-toxic anti-cancer medications, which eventually result in the death of cancer cells. Further investigation has revealed that the flavonoid kaempferol causes DNA to break, which in turn stimulates the production of p53 and its subsequent phosphorylation, which in turn eliminates the proliferative signals that breast cancer cells utilize to proliferate themselves. Theaflavins, which are the most bioactive of all the polyphenols found in tea, make up the vast bulk of this beverage. Rheumatoid arthritis is a form of inflammatory joint disease that can lead to a loss of cartilage and bone in joints that are affected by the condition.

There is some evidence to suggest that theaflavins, which are a kind of phytonutrient, may help to preserve healthy cartilage that also has a youthful look. Because it promotes the eating of a wide variety of fruits, vegetables, and grains, in addition to a large variety of nuts and seeds, the Mediterranean diet is high in monounsaturated and omega-3 polyunsaturated fats (Ferguson, 2016). These fats are found in abundance in the Mediterranean diet. It is possible for nutrients to either activate or deactivate genes in their bodies.

Researchers used rat models and investigations done in cell culture to study how widespread gene expression and disease impact food intake. Specifically, the researchers wanted to learn how widespread gene expression and illness influence food consumption. Alterations in DNA methylation have been linked to various disorders whose incidence is rising at an alarming rate. To methylate DNA, it is essential to convert nutrients that are rich in methyl donors into fuel. This transformation is called methylation. The increased hunger and resultant weight gain of mice may be traced back to a gene in the species known as agouti. People had an increased likelihood of acquiring diabetes as well as cancer since this gene was present in their bodies.

2.2 DIET AND GENETICS

Variables in the diet and the broad spectrum of genetic differences:

The successful conclusion of the Human Genome Project has been of tremendous assistance to researchers working in the field of nutritional genomics. With the use of this information, researchers have been able to establish several connections between certain genes, foods, and diseases. The decoding of the human genome has made it clear that the genetic make-up of people living in different human cultures is very diverse from one another. Diet is believed to have played a role in the formation of hundreds of thousands of single nucleotide polymorphisms (Fenech, 2008).

Single-nucleotide polymorphisms in genes that are involved in the metabolism of drugs, environmental contaminants, or dietary components can have a significant influence on the exposure-response of an individual, including nutrition. This is because SNPs are located in regions of the gene that are known to be highly conserved.

There is a complex web of interaction that exists between a person's genes and the foods they eat.

To keep the body in its homeostatic condition, it is necessary for the expression of genes, the metabolic rate, and the diet all to operate together. Researchers have demonstrated that illnesses that are associated with nutrition or are dependent on nutrition are caused not by a single gene but rather by a combination of nutrients and numerous genes. The majority of the individual variation in response to nutrition may be linked to genetic diversity. Nutrigenetics is the study of the influence that genetic variation has on gene expression, as well as the discovery of genetic variants as risk factors for human disorders that are connected to diet or nutrition in general (Fallaize et al., 2013). Additionally, nutrigenetics is the study of the influence that genetic variation has on gene expression.

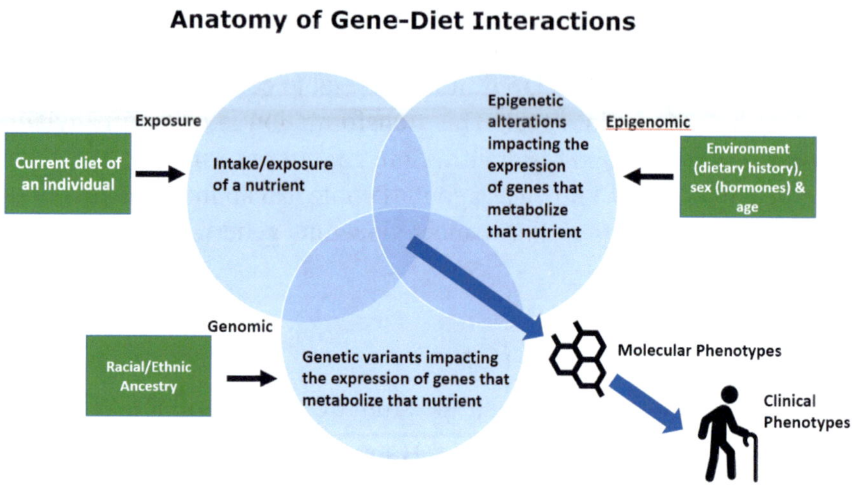

Figure 2.4: Nutritional genomics.

Source: Mullins VA, Bresette W, Johnstone L, Hallmark B, Chilton FH. Genomics in Personalized Nutrition: Can You "Eat for Your Genes"? Nutrients. 2020; 12(10):3118. https://doi.org/10.3390/nu12103118

The field of study known as nutritional genomics examines how one's genes interact with their food intake. Diet and human DNA interact in a way that allows information to flow in both directions. It exerts a regulatory and marking influence on the metabolic response as well as on the expression of genes. After that, it may influence the person's health and the likelihood that they will become ill. An individual's nutritional status, metabolic reactivity, and vulnerability to diet-related or connected health problems can all be influenced by their ancestry in some way. Ancestry is defined as "the history of a people." Second, the consumption of food influences transcription factors, which in turn affect the expression of genes and, as a consequence, the metabolic processes of molecules (Elsamanoudy et al., 2016). As a consequence of the relationship between human food and the human genome, new subfields of research have evolved, such as nutrigenetics and nutrigenomics.

It has been hypothesized that the area of study known as nutrigenetics focuses on the process by which genetic variants define an individual's illness risk, nutrient daily needs, cellular metabolic response, and behavior towards bioactive food components or nutritional treatment. The purpose of nutrigenetics is, first and foremost, to shed light on the influence that the variety of genes has on the connection between nutrients and illnesses.

Nutrigenomics is a term that refers to the study and investigation of the effect that a person's diet has on their genome. It is interested in the functional influence that diverse dietary components have on the genome, transcriptome, proteome, and metabolome, among other things.

Both nutrigenomics and nutrigenetics center on the investigation of the relationship that exists between a person's eating habits and their genes. On the other hand, the processes of interaction between nutrition and genes that determine the risk of having a disease are quite diverse from one another. This is because the likelihood of acquiring a disease is determined by the interaction between nutrition and genes. The study of how different nutritional components influence gene expression is another application of the term "nutrigenomics." These nutrient components could comprise agents based on food, agents that restrict dietary intake, or agents that augment nutritional intake. The effect that nutrition has on the genome-wide transcriptome, proteome, and metabolome of cells, organs, and species will become increasingly obvious as a result of this. In addition to this, it will find genes that impact the risk of developing illnesses related to nutrition, and it will confirm these genes (Dimitrov et al., 2016). This may also aid researchers in discovering how metabolic pathways are regulated by foods and how these controls can be turned off in the early stages of diet-related and diet-dependent illnesses.

Dolinoy and Jirtle advocated the establishment of a dietary signature to describe the changes in gene expression, protein expression, and metabolite synthesis that are brought on by nutrition. This was done so that the differences could be identified and characterized. As a consequence of this, the goal of nutrigenomics is to ascertain how the characteristics of food influence the function of cells as well as the equilibrium of tissues located throughout the interior environment of the body.

2.3 THE FUNDAMENTALS OF GENETICS APPLIED TO NUTRITION

Four distinct ideas provide the foundation for the study of dietary genomics as a discipline. It is a commonly held belief that nutrition has a substantial role in the development of several different diseases, particularly in the cases of specific individuals and under certain conditions. Dietary nutrients can modify the structure of genes as well as the expression of genes, and as a consequence, they have the potential to transform the human genome. The one-of-a-kind genetic variety of each person is responsible for the delicate balancing act that exists between health and sickness (Dimitrov, 2011). There is a possibility that one's dietary habits can affect gene regulation, which in turn may play a role in the early stages, the development, and progression of chronic illnesses.

Several Illustrations from the Field of Nutrigenomics:

The consumption of foods high in cholesterol inhibits the transcription of the gene encoding the enzyme hydroxymethylglutaryl-CoA reductase. Polyunsaturated fatty acids in the diet reduce the quantity of mRNA in hepatocytes, which in turn lowers the amount of mRNA that is responsible for lipogenic enzymes. This results in an inhibition of the generation of fatty acid synthase mRNA. There is a correlation between the degree of unsaturation of the fatty acids that are used in this procedure and the outcome.

One mutation in a single gene is responsible for the condition known as phenylketonuria. People who have phenylketonuria should steer clear of foods that contain a high concentration of phenylalanine. In several Asian ethnic groups, the enzyme aldehyde dehydrogenase is present in very low concentrations. The breakdown of ethanol into its parts is the job of this enzyme (Dall'Asta et al., 2022). Those who are afflicted feel something unpleasant after they have drunk alcohol as a result of this condition. Galactosemia is brought on by a genetic defect in one of the three enzymes that are necessary for the process of galactose metabolism. Galactosemia is a genetic disorder.

Several Illustrations from the Study of Nutrigenetics

The methylenetetrahydrofolate reductase gene is an example of a gene-nutrient relationship that may be specified in further depth. Both the metabolism of folic acid and the control of homocysteine levels in the blood is under the control of the MTHFR gene, which plays an essential part in both processes. There is a correlation between a specific SNP in the MTHFR gene and increased levels of homocysteine in the blood of carriers, in particular in conditions in which folic acid is insufficient in the diet. This is especially true in situations where there is a lack of folic acid in the diet.

Increased levels of homocysteine have been linked to an increased risk of colon cancer and cardiovascular disease, in particular when dietary folic acid levels are considerably below the minimum daily requirement. This is the case when levels of homocysteine are considerably below the minimum daily requirement.

The study of how genes are handed down from one generation to the next falls within the purview of the scientific discipline known as nutritional epigenomics. Numerous studies on the subject of epigenetics have focused on inherited variations in DNA and, as a result, protein synthesis, as well as the interaction between DNA and histones, which can result in changes in chromatin structure without altering the nucleotide sequence. These studies have been conducted on a variety of organisms, including humans, mice, and flies (Dang et al., 2014). These studies have been carried out on a wide range of organisms, one of which was the human population. Epigenetic processing, such as cytosine methylation and histone acetylation, as well as microRNAs and non-coding RNAs, which are also implicated in the epigenetic setting and regulation of the process of gene expression, can therefore be the cause of changes in gene expression rather than changes in DNA sequence. This is because epigenetic processing can regulate gene expression in a way that DNA sequence changes cannot. This is the case because epigenetic processing can take place regardless of whether or not there are modifications to the DNA sequence. The genome of a mammalian organism includes the genetic code for all of the many forms of microRNAs. To control the expression of genes, these miRNAs make modifications to the mRNAs that they target. In yet another part of the study, it was shown that the process of chromatin modification involves the participation of miRNAs. These findings suggest that microRNAs, DNA methylation, and histone changes may all act together to influence gene expression.

Nutrients have a role in epigenetic changes, which are what regulate and control the signaling pathways that are located inside cells. These epigenetic alterations may be controlled by nutrients. The Dutch Famine Cohort provided the first evidence of the involvement of nutritional situation in epigenetic modifications, as well as the mechanism by which these epigenetic abnormalities affect the health effect on long-dated amplitude. The Dutch Famine Cohort was comprised of people who lived in the Netherlands during the Dutch Famine of the 1840s. In addition, the Dutch Famine Cohort was the first cohort to study the link between one's dietary circumstances and the epigenetic changes that result from it. According to the findings of these cohort studies, limiting one's calorie intake while one is pregnant may raise the risk that one's kids would have metabolic disorders (Deb et al., 2014). This possibility was shown to exist even when the pregnant woman did not restrict her caloric intake. It was discovered that women who were hungry throughout the first trimester of their pregnancies had an increased chance of being unwell in their early adult years. This was the case even if the women went on to have healthy pregnancies. In the course of its investigation, the Dutch Famine Cohort Study came upon two noteworthy discoveries. Two important points: first, a person's diet can both cause and result in epigenetic abnormalities during a critical stage of embryonic development; second, the epigenetic alterations caused by a person's diet are almost certainly passed down to subsequent generations. Both of these points are related to the concept of epigenetic inheritance.

Figure 2.5: DNA chip containing the genome of an obese person.

Source: https://hydrobot.com/about/dna-microarray-chips/

2.4 CANCER RISK AND DIET

Enzymes that catalyze biotransformation are responsible for the digestion of the great majority of the cancer-preventative, cancer-causing, and mutation-causing chemicals that are present in the diet. DNA polymorphisms that alter the activity of enzymes or the production of proteins can have an impact on the likelihood of an individual getting cancer. It is believed that there are approximately 25,000 distinct bioactive dietary components that might be identified in human food (Connolly & Dawson, 2015).

Over five hundred different bioactive food components have been singled out as possible contributors to the development of cancer. Some of these factors do have a role in the development of cancer, whereas some of these factors do not. It is hypothesized that people who are genetically predisposed to developing cancer will have a lower risk of developing the disease if they consume a diet that is high in carcinogens, mutagens, and cancer-preventing micronutrients. This is because individuals who are genetically predisposed to developing cancer will have a lower risk of developing the disease.

These bioactive dietary components can come from a broad number of different sources, including animals and plants, amongst other potential origins. These facets of dietary habits have been hypothesized to reduce the chance of acquiring cancer in some people. They can be classified as either carcinogenic or anticarcinogenic based on how they interact with certain genetic variations and how they impact one another (Corella & Ordovas, 2009). Interactions that take place in the time in between meals have the potential to influence the reaction as a whole.

Molecular targets of foods that include compounds with biological activity

The tumor affects the normal cellular activity, the neoplastic transformation of normal cells into malignant cells, and the biological behavior and property of the tumor, and all of these things are influenced by the tumor itself. Bioactive dietary nutrients and components affect the normal cellular activity of the body. Even though the pathophysiological process differs based on the kind of food that is eaten, it is thought that these three phases have a substantial influence on behavior and the risk of acquiring cancer. This is the case even though the pathophysiological process varies. Numerous studies have demonstrated that every diet influences the enzymes that are involved in the xenobiotic metabolism at both the Phase I and Phase II levels. This is proof that normal cell function may be maintained.

It has been demonstrated that certain components of the diet can lower the chance of developing cancer through a variety of distinct ways, one of the key mechanisms being a shift in the metabolism of carcinogens. Nuclear receptors that sense xenobiotics, including AhR, PXR, and RXR, are crucial for regulating gene expression in enzymes that are involved in Phase I of the xenobiotic metabolic pathway (Chen & Zhao, 2013). The great majority of carcinogens are activated by these receptors, and this is how they cause cancer. Phase II enzymes are responsible for catalyzing the conjugation of these carcinogens, which are transcriptionally controlled in a variety of ways via different signaling pathways.

There is evidence to suggest that reactions to dietary components that have a role in the prevention of cancer are related to multiple enzymes that are digested and changed. This is the case because there is evidence to suggest that these responses are associated with cancer prevention. Garlic, for example, has been linked to the autoactivation of CYP2E1, although this mechanism does not cause garlic to have any influence on any of the other isozymes in the CYP450 family. In addition, genetic polymorphisms in the regulatory areas of metabolizing enzymes and transporter proteins, such as AhR and PXR, could influence how the body responds as a whole to the bioactive components of the food.

Previous studies have demonstrated that a person's diet may influence the rate of advancement of neoplastic growth as well as the programmed death of cells in the body. These effects can be caused by a variety of dietary factors. Multiple protein kinase complexes, each of which is formed of cyclin and a molecule of cyclin-dependent kinase, are accountable for the regulation of crucial processes that take place during the cell cycle. In addition, the specific mix of components that constitute the diet influences the particular stages of the cell cycle that are being discussed here (Carlberg, 2019). It has been established that nutrients in the diet, both necessary and non-essential foods, can shift and impact cell cycle checkpoints and that this ability gives them the potential to play a role in the growth of tumors as well as the repression of proliferation.

Compounds such as apigenin, curcumin, epigallocatechin-3-gallate, resveratrol, and genistein, in addition to allyl sulfur, have been associated with causing a disturbance in the cell cycle. Some of these alterations may be caused by the processing of freshly generated proteins during the post-translational stage. As a consequence of this alteration, the phosphorylation patterns of the factors that play a critical role in maintaining control over cell division are altered.

According to one school of thought, the activity of tumors may be altered by other aspects of a person's diet by accelerating the process of cell death and boosting apoptosis. This is how the hypothesis works. There are two different kinds of mechanisms that are involved in the process of apoptosis: the intrinsic process, which is handled by mitochondria, and the extrinsic process, which is mediated by death receptors. Both of these processes are referred to as death receptors. Dietary components have the power to either speed up or slow down the advancement of the apoptotic cascade. This is because they can modulate the transcription of mRNA, the expression of proteins, and their functions (Castro et al., 2006). Some of these bioactive components of food may induce cells to commit suicide by triggering an increase in the creation and synthesis of reactive oxygen/nitrogen species (ROS) within the cell. This may happen when the cell is exposed to an excessive amount of the bioactive component.

2.5 EVALUATING DIETARY INTAKE

To investigate and determine the relationship between dietary habits and the risk of developing cancer, it is necessary to have precise and cost-effective methods for monitoring the consumption of specific essential and non-essential bioactive nutritional components. This is necessary to monitor the consumption of specific essential and non-essential bioactive nutritional components. This is important to research and ascertain the nature of the relationship. Even though both the food frequency surveys and the 24-hour recall methods of data collection have significant limitations, they are still the most common methods of data collection used to research dietary exposures (Bionaz et al., 2015).

It is necessary to develop new methods for measuring these parameters and their effects at the cellular and molecular levels. This is because absorption, metabolism, tissue distribution, and excretion all affect the amount of these bioactive dietary ingredients that reach the target sites. The combination of measures of dietary intake coupled with analyses of bioactive food components or metabolite concentrations in tissues and/or bodily fluids might reveal unique insights into individual food reactivity as well as the possibilities for long-term dietary exposures.

2.6 THE LINK BETWEEN NUTRITION, GENETICS, AND CANCER AND THE VARIABLES THAT IMPACT IT

2.6.1 Variations in Genes between Humans

Because of the heterogeneity between individuals in terms of both the structure and sequencing of their genomes, it is tough and confusing to comprehend how genes interact with nutrition. The single nucleotide polymorphisms (SNPs) and copy number variations (CNVs) that are most abundant in the human genome are the structural modifications that can contribute to nutritional genetic diversity (Brennan & de Roos, 2021). It has been noted that there is variation in the copy number of the -amylase gene as well as numerous cytochrome P450 genes. A higher copy number variant of the enzyme -amylase has been associated with a history of consuming meals high in starch, which has caused health concerns.

2.6.2 Epigenetics Influences Dietary Response

Within the context of the study of epigenetics, the term "epigenetics" refers to the ongoing modification of DNA structure. These modifications have the potential to affect the expression of genetic information without in any way altering the sequence of the information itself. In the subject of epigenetics, the bulk of the study and attention has been concentrated on the process of chromatin remodeling, which takes place as a consequence of the alteration of histones and the methylation of DNA.

Changes to the epigenome of an individual have the potential to be involved in the pathophysiology and development of a variety of illnesses, including cancer. One of these disorders is cancer. Epigenetic change of the relevant CpG islands of genes involved in cell cycle control, DNA repair, angiogenesis, and apoptosis can block and deactivate those genes, as revealed by a large number of studies.

The important groups of regulatory genes that are influenced by DNA hypermethylation include cyclin-dependent kinases, phosphatase, insulin-like growth factor-II, E-cadherin, and glutathione-S-transferase. DNA hypermethylation also affects the expression of genes that code for glutathione-S-transferase. The expression of glutathione-S-transferase can be altered as a result of DNA hypermethylation as well. It has been established that certain dietary components and/or combinations of food components, such as vitamin A and zinc, may alter DNA methylation patterns (Braicu et al., 2017). Other examples of food components that can influence DNA methylation patterns include: It is also feasible that dietary components that aren't vital to survival are playing a role in this process.

The discovery that diets low in methyl can drastically affect the patterns of methylation that are found during the transformation of normal cells into malignant ones were made by Fang in the year 2007. Because of its ability to exert a direct regulatory influence on the DNA methyltransferase enzyme, genistein and other similar soy isoflavones be capable of reactivating genes that have been inactive as a result of methylation. The chemical known as genistein is classified as a phytoestrogen isoflavone. This classification indicates that genistein may be found in plants and has a structure that is analogous to that of estrogen.

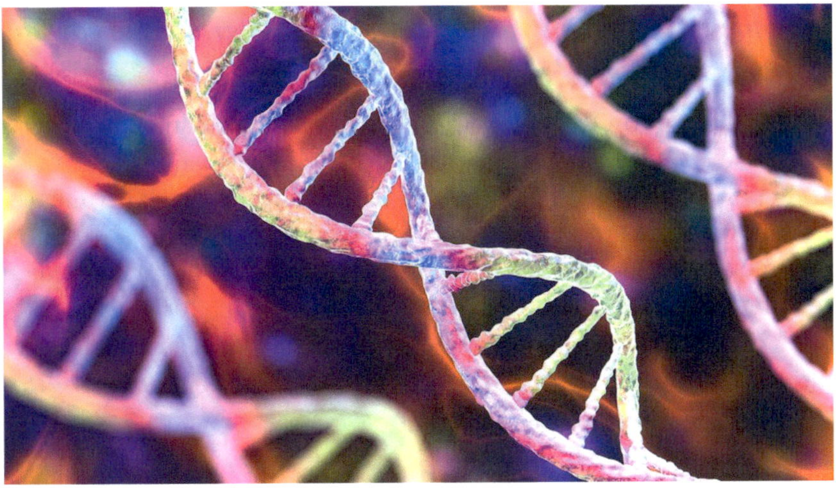

Figure 2.6: Human genome Image.

Source: https://www.bbc.com/news/health-60952034

Gene expression can be affected by modifications to histones, which can be either permanent or temporary depending on the circumstances. These include things like reversible histone acetylation, altering the pattern of methylation, phosphorylation and dephosphorylation, ubiquitination, and biotinylation. The modulation of histone deacetylase is an essential component in the process of influencing the behavior of malignancies (Bragazzi, 2013).

In a study that was carried out *in vitro*, several other components of the diet, such as butyrate, diallyl disulfide, and sulforaphane, behaved in a manner that was consistent with that of weak ligands for the enzyme histone deacetylase, thereby influencing the activity of the enzyme. The behavior of these other components of diet was determined by observing how they interacted with the enzyme. It has been demonstrated that the isothiocyanate compound known as sulforaphane, which can be found in several vegetables that are high in antioxidants, can increase the level of histone acetylation both globally and locally in the promoter regions of the P21 and Bax genes. This effect has been shown to have a positive impact on cancer prevention and treatment. These genes have been linked to cellular senescence as well as genes involved in the process of apoptosis.

How much of it was consumed during what period and for how long was it consumed?

The total reaction to meals or nutritional supplements can be significantly influenced by several important factors, including the timing of nutrient consumption as well as the length of that intake. In an experiment that Virgili and Perozzi carried out, they discovered that the amount of time that rats were exposed to genistein in their food was a key factor in predicting whether or not the rats got breast cancer (Benítez et al., 2017).

The genes that cause cancer as well as the dietary components that are connected with those genes

It is well known that the likelihood of a person developing cancer varies substantially from person to person, and this is something that should not be surprising. Individual variations in the risk of acquiring cancer can be partially attributed to polymorphisms that are present in the human genome. Numerous research articles have compared the prevalence of various gene polymorphic forms among cancer patients to the frequency of the same gene polymorphism among healthy, unaffected controls. These investigations have found that cancer patients had a higher incidence of certain gene polymorphic forms. These investigations have been carried out on a wide array of distinct types of cancer.

Researchers have focused the majority of their emphasis on the polymorphisms of carcinogenic gene variations. These polymorphisms modify the bioavailability, metabolism, affinity, and activity of a wide variety of dietary components (Wiwanitkit, 2012). In addition, studies have been carried out on the genes that are answerable for the synthesis of sex hormones as well as the genes that are engaged in the replication and repair of DNA.

Genes that are linked to the development of cancer are referred to as carcinogenes. These genes are also referred to as susceptibility genes. The term "candidate gene" refers to a gene that has been found in an area of the chromosome that is believed to play a function in the presentation of a disease. It is also feasible to locate a potential gene by basing the search on the position of the gene near a certain genomic region as well as the gene's correlation with a particular quality.

Carcinogenic genes are those genes that are involved in the metabolic pathways of carcinogens.

During stages I and II of the metabolic pathway, xenobiotic-metabolizing enzymes are what's responsible for the biotransformation and/or detoxification of a foreign chemical or carcinogen. This procedure's objective is to transform potentially harmful compounds that are water-insoluble and difficult to excrete into ones that are water-soluble and easy to get rid of. The product of the Phase I reaction is conjugated to a transporter so that it may be removed in urine or bile. This makes it feasible for the substance to be excreted. The addition of one or more hydroxyl groups to molecules by phase I enzymes changes the molecules into hydrophilic polar intermediates that are capable of attaching to a specific conjugating substrate and possibly increasing their reactivity to combine with DNA or protein (Wittwer et al., 2011). These intermediates are created when molecules undergo the addition of one or more hydroxyl groups. Oxidoreductases are the category that Phase I enzymes fall under. This binding to DNA can occasionally result in the production of carcinogens because it speeds up the process of promutagenesis.

A group of genes referred to as the CYP gene family are responsible for encoding the phase I enzymes known as cytochrome P450. Phase II enzymes, also known as conjugating enzymes, attach themselves to conjugating substances such as glutathione and glutamine to eliminate polar intermediates from the body. Phase II enzymes are also known as conjugating enzymes. In some contexts, phase II enzymes may also be referred to as oxidoreductase enzymes. Researchers have identified a variety of distinct enzyme families that play a role in the process of Phase II metabolization. This category includes enzymes such as glutathione-S-transferases, N-acetyltransferases, and UDP-glucoronosyl-transferases, to name just a few examples. Depending on the particular enzymatic activity that is either stimulated or suppressed, as well as the substrate that is at play, polymorphisms in the genes that code for Phase I and Phase II enzymes can either increase or decrease the likelihood of developing cancer. This is because the likelihood of developing cancer depends on the particular enzymatic activity that is either stimulated or suppressed.

Cytochrome P450 is a family of enzymes that catalyzes the oxidative transformation of a wide variety of chemicals, including those that are created within the body as well as those that arrive from the outside. Cytochrome P450 enzymes are found in almost all cells (Van Breda et al., 2015). They appear at different times and in different ways at different rates of speed depending on where in the body they are placed. Because of this, the role that these enzymes play in gaining knowledge of the genetic vulnerability, the metabolism of carcinogens, and the metabolism of chemopreventive medicines is of the utmost significance.

Genes that contribute to the production of steroid hormones, as well as their destruction, are involved.

Estrogens are the primary factor in the development of breast cancer because of their ability to trigger gene expression in a manner that is associated with the growth and multiplication of cells. The production and metabolism of estrogen is a complicated process that is impacted by a wide variety of genetic differences. There is a possibility that variations in the genes that are a part of the synthesis pathway might affect estrogen production. CYP17 and CYP19 are two genes that are thought to be involved in this pathway. Changes in estrogen bioavailability, half-life, and activity can be brought about by other metabolizing enzymes, such as the catechol-O-methyltransferase enzyme, which is involved in the biotransformation of estrogen.

The 1931T-C polymorphism in the CYP17 gene promoter region has been the subject of research and characterization efforts. The chemical reaction that turns pregnenolone into 17-hydroxypregnenolone is catalyzed by CYP17. The homozygous version of this gene is rather common, being present in around 13 percent of the total population. Multiple studies have found a correlation between certain versions of the CYP17 polymorphism and an increased likelihood of developing breast cancer (Su & Lu, 2012).

The COMT enzyme is a Phase II metabolizing enzyme that is involved in the inactivation of catechol estrogen molecules by conjugating them with a methyl group. This is done so that the molecules can no longer function as estrogens. A single change in amino acid sequence is responsible for several variations found in the COMT gene. This alteration is associated with a delay in the methylation process, which ultimately leads to increased estrogen levels. Women who possess the COMT Val158Met polymorphism have been shown in several studies to have a higher risk of developing breast cancer, both in terms of its incidence and its prevalence.

2.7 REPAIRING DNA GENES

The process of repairing damage to DNA, which can be caused by either endogenous or exogenous causes, is referred to as DNA repair. The genes that repair DNA protect the genome from being altered in any way. Polymorphisms in the genes that code for the DNA repair system can affect the ability of those genes to prevent mutation. Polymorphisms in those genes have been linked to alterations in the enzymatic activity of the protein product as well as changes in the capacity of the protein product to contact its protein or receptor partner.

The X-ray cross-complementing Group 1 protein is perhaps the most well-known example of a DNA repair process. This is a part of the system that is often referred to as the base excision mechanism. Three different genetic variants may be found in the coding area of the XRCC1 gene (Alagawany et al., 2022). These variants can be found at codons 194, 280, and 399. These genetic variations encode amino acid changes that have the potential to affect the functioning of XRCC1. Polymorphisms in XRCC1 codons 194 and 399 are related to an increased chance of developing oropharyngeal squamous cell carcinoma, particularly in heavy smokers and drinkers. These findings have been gleaned from several different investigations. In addition, it has been linked to colorectal cancer in Egyptians, although the variant known as 399Gln has been attributed to a decreased risk of skin cancer.

In nucleotide excision repair, the DNA helicase enzyme that is encoded by the xeroderma pigmentosum group D gene plays an important role. Two different genetic variations may be identified in the XPD gene. These variants are called Lys751Gln and Asp312Asn. These two differences in genetic polymorphism have been linked to a reduced capacity for repairing nucleotide excision. Carriers of the Asp312 XPD gene had almost double the risk of developing lung cancer compared to individuals who did not have the gene (Singh & Sharma, 2019). On the other hand, several studies have demonstrated the existence of a connection between DNA damage and the 751Gln XPD genotype in people who are in otherwise good health.

2.8 CONCLUSION

Cancer is the result of a complicated interaction between genetic and environmental factors, which ultimately results in the disease's genesis and progression. Some researchers have hypothesized that the consumption of food and the use of nutritional supplements, which are both considered to be major environmental variables, could have a role in the progression of cancer. In addition, several studies have pointed to a significant connection between the quantity and quality of the nutrients in one's diet, as well as the incidence and genesis of cancer.

Even among those who consume the same food, there is a significant difference in the likelihood of developing cancer. This might be due to variations in their genetic polymorphisms, which eventually led to the development of the concepts of nutrigenomics and nutrigenetics. Nutrigenomics and nutrigenetics may explain the connection between certain dietary habits and the influence of genetic variations on the development of cancer (Sikalidis, 2019). Nutrigenomics is finding more and more uses, including the study of the origin and cause of cancer, the evaluation of treatment responses, and the investigation of disorders that are connected to or dependent on nutrition. After that, nutrigenomics may be used to successfully treat cancer through the creation of individualized dietary recommendations that are based on an individual's genetic profile.

CHAPTER 3

THE INTERACTION OF FOOD AND GENES

CONTENTS

3.1 INTRODUCTION

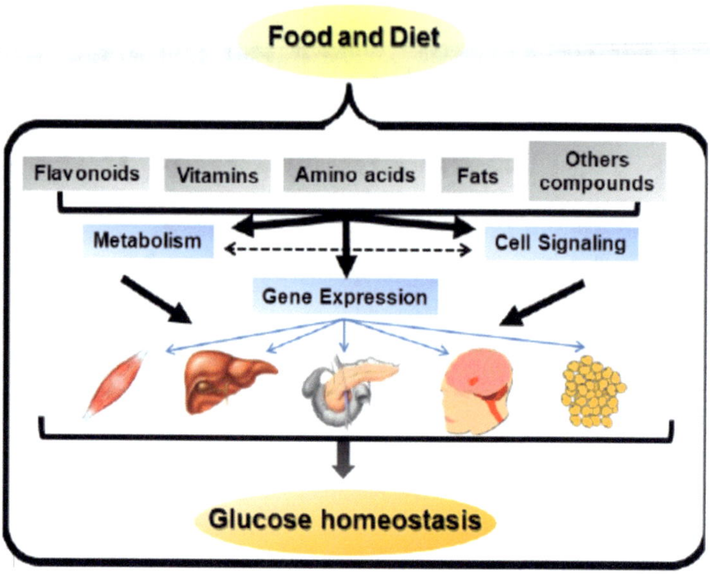

Figure 3.1: The interaction of food and genes.

Source: Berná, Genoveva & Oliveras-López, María & Jurado-Ruiz, Enrique & Tejedo, Juan & Bedoya, Francisco & Soria, Bernat & Martín, Franz. (2014). Nutrigenetics and Nutrigenomics Insights into Diabetes Etiopathogenesis. Nutrients. 6. 5338-5369. 10.3390/nu6115338.

Recent advances in the study of nutrition have resulted in the creation of high-throughput techniques for the fields of genomics, proteomics, and metabolomics. These methods have the potential to make it much less difficult to research the potential nutritional value of foods and other aspects of diets. As a direct consequence of this finding, researchers will be able to identify the types of foods, individuals, and social groups that interact with particular genes. Recent studies in the field of nutrition have provided evidence for the significance of the role that nutrients play in the expression and regulation of genes (Riscuta, 2016). These studies have shed light on the significance that nutrients play in the processes of expression and regulation of genes. The fact that the expression of genes in prokaryotic organisms can change in response to changes in the nutritional state of the organism is a fact that is well known. Gene expression variability is the term used to describe this phenomenon.

When compared to single-celled organisms, multicellular animals need a more complex combination of hormones, neurons, and nutrients in order to exert control over the expression ofDNA. This is because multicellular animals have more than one hormone receptor. This particular organism consists of multiple cells, making it distinct from single-celled organisms. It is a fact that is rarely acknowledged, although it is a reality, the nutrients that animals consume have the potential to change their DNA. The genetic make-up of each individual is one-of-a-kind, and the expression of each gene in a person's body is controlled by a different aspect of the diet that person consumes.

Nutritional genomics is a branch of research that investigates how the consumption of food that is rich in a variety of nutrients can have an effect on the cellular expression of genetic information within the body of an organism. The production of proteins is a byproduct of the process that occurs when genes are in their active state and go through activation or expression. Scientists have discovered that the fields of molecular biology and genomics have been of tremendous assistance to them in their search for the genes that are accountable for the production of nutritionally significant proteins (Riscuta & Dumitrescu, 2012). This search has been facilitated by the disciplines of genomics and molecular biology. These proteins include digestive enzymes and transport molecules, which are responsible for transporting nutrients and cofactors to the sites where they will be utilized. They are also responsible for breaking down food into its parts. These proteins are necessary for living things.

The researchers made use of a variety of methodologies in the course of their investigation. The patterns of gene expression are a reflection of an organism's physical qualities, such as the color of its hair, its body weight, and whether or not it is afflicted is determined by the genes. Examples of these traits include: It is not always the case that an individual's genes are responsible for creating their phenotype in the manner that they seem to the outside world. This is not necessarily the case in every circumstance. Obesity, the various forms of cholesterol (LDL and HDL), homocysteine levels, and cholesterol levels themselves are all dietary variables that have the ability to impact one's cholesterol levels and the implications that cholesterol may have (the genetic component). Another example of a dietary element that has the ability to alter cholesterol levels is homocysteine levels. These levels can be affected by dietary factors.

3.2 FOOD NUTRITION AND GENE REGULATION

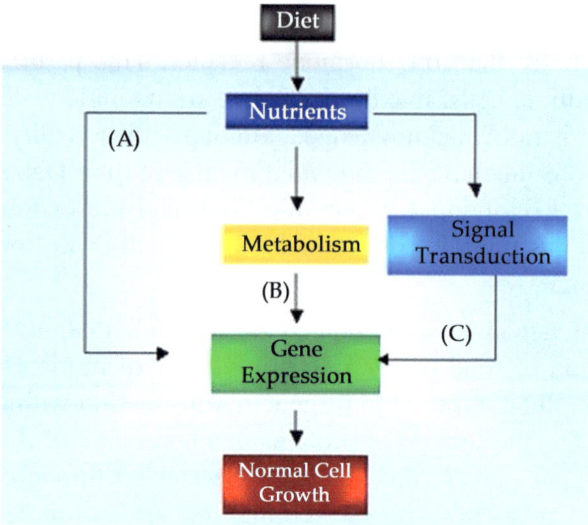

Figure 3.2: Food Nutrition and Gene Regulation.

Source: http://foodexposed.co.za/wp-content/uploads/2015/05/dietarychemicals_Pathways.jpg

Components of the diet have an effect on every stage of the process that leads from a gene to a protein or metabolic product. This includes the signal transduction and post-translational changes that take place along the way. It is now possible, thanks to advances in genomic technology such as nutrigenomics, transcriptomics, metabolomics, and epigenomics, to modify the carcinogenesis process through dietary means. This process is referred to as the nutritional regulation of the route. It is now possible, thanks to the utilization of nutritional genomics, proteomics, and metabolomics in conjunction with one another, to investigate the influence that components of diet have on the operation of cells and the expression of genes. Because of the interaction that nutrients have with transcription factors, dietary habits can have an effect on the expression of genes (Ray, 2014).

The nuclear receptor superfamily of transcription factors is a significant subcategory of nutrition detectors and is considered to be one of the most important. The genome of a human being contains 48 different members of this particular family. This superfamily of receptors is responsible for the binding of a variety of nutrients and metabolites, and it contains several receptors that are responsible for this process. The activation of many genes is caused by interactions between nuclear receptors and particular nucleotide sequences in the promoter regions of many genes. These sequences are known as response elements, and they are located in the promoter regions of many genes. To give one example, the PPAR is associated with the process of metabolizing lipids, whereas the LXR is associated with the process of metabolizing cholesterol compounds.

There is a possibility that the conformation of nuclear receptors will change as a consequence of the interaction between ligands and nuclear receptors. They have the effect of resulting in the dissociation of repressors and the recruitment of coactivators, both of which are necessary for the activation of transcription. They also have the effect of resulting in the activation of transcription. These transcription factors regulate the amount of DNA transcription in response to variations in food intake, which allows them to function as nutrition sensors in metabolically active organs such as the liver, the stomach, and adipose tissue (Qadeer & Ashhar, 2014). The functions of nuclear receptors include, but are not limited to, the metabolism of food, the development of embryos, the proliferation of cells, and the differentiation of cells. However, these functions are not the only ones that nuclear receptors perform. In addition, nuclear receptors play a part in the process by which cells differentiate into different types.

Despite having a very short genome, *E. coli* is able to differentiate between extremely similar carbohydrates, such as lactose and glucose. This is due to the fact that *E. coli* is so small. Although the lac operon and a huge number of other regulatory genes are principally responsible for this, there are also a significant number of other genes that are involved. In *E. coli* and other intestinal bacteria, the lactose transport and metabolism can be controlled by the lac operon and the transcription regulatory region. The lac operon is comprised of three different genes altogether. The lac I gene is the only gene that can be found in the lac transcription regulator region (encoded by the repressor gene). The lac operon is comprised of three different genes: lacZ, lacY, and lacA (lac ZYA). The gene known as LacZ is the one that is in charge of encoding galactosidase for the cell to utilize. Lac permease and the enzyme thiogalactoside transacetylase are controlled by the genes LacY and LacA, respectively.

Lactose permease is referred to as an enzyme that transports lactose from the extracellular environment into the cell. Following this step, the cytoplasmic enzyme galactosidase is responsible for further degrading the lactose molecule into glucose and galactose. If the cell does not have access to lactose or another source of sugar, such as glucose, it will not be able to manufacture these enzymes. It is possible to accomplish this goal by employing a lactose regulatory control for lac genes that is reliant on lactose and exclusive to lactose. When the LacZYA genes are cotranscribed, the result is the formation of a polycistronic mRNA molecule. The beginning of the transcription process in all of the genes that are located upstream of the target gene is initiated when RNA polymerase binds to the promoter region (Pauletto et al., 2020). There is a connection between the promoter and the cAMP-linked catabolic activator protein. The levels of cAMP decrease if there is a high quantity of glucose within the cell. According to the findings of this research, it would appear that the enzyme known as adenylate cyclase, which is responsible for the production of camp, has the capability of determining the intracellular concentration of a glucose metabolic intermediate.

The formation of a complex including cAMP and the cAMP receptor protein occurs both when glucose levels fall and when cAMP levels rise. This complex is produced when glucose levels fall. As a consequence of this change, there was an increase in affinity for the lac operon, which is located in close proximity to an RNA polymerase binding site. This connection results in the formation of a closed promoter complex, which in turn causes the lac operon to be activated to start the transcription process. The products of all three genes, lacZYA, are subsequently converted into messenger RNA by an enzyme called RNA polymerase. The lac I gene, which is always expressed (constitutive) and is situated close to the lac operon, is responsible for the production of repressor proteins. This is because the repressor attaches to the lac operator, a short DNA sequence that is located just downstream of the promoter and just ahead of the start sequence of the lacZ gene. This prevents the lacZ gene from being generated if the medium does not contain lactose. RNA polymerase is unable to form a connection with the promoter because it is linked to the repressor, which is also linked to the operator.

As a consequence of this, transcription is either not feasible at all or of very low quality. When lactose is present in the culture media, its metabolite, allolactose, reaches out and forms a binding relationship with the repressor. As a consequence of their interaction with allolactose, repressors are unable to successfully form stable complexes with the operator. As a consequence of this, RNA polymerase is capable of successfully transcribing lac genes, which increases protein output. Another regulatory mechanism is activated whenever glucose enters the cell through the PEP-dependent phosphotransferase system (Peña-Romero et al., 2018). This procedure takes place on the inside of the cell. It is believed that EIIBGlc, in combination with glucose transport, phosphorylates glucose, therefore inhibiting lac permease (lacY) from transferring lactose into the cell. This occurs when glucose is transported. During the process of glucose transport, glucose phosphorylation will take place. The passage of the lac operon inducer is impeded by conditions in which glucose and lactose are present in the environment. This occurrence is referred to as "inducer exclusion," which is a word used to characterize it.

3.3 EPIGENETICS AND NUTRITION

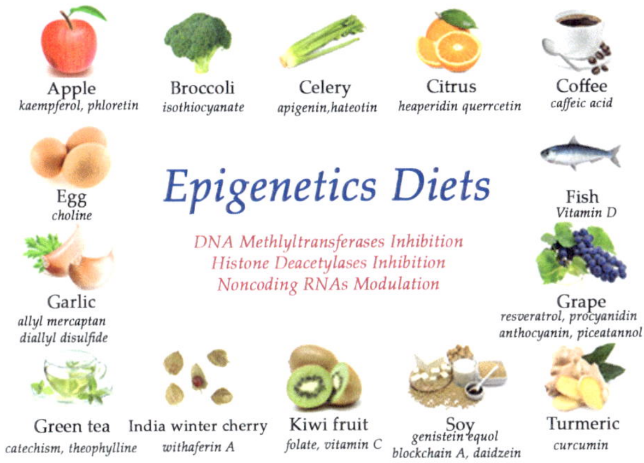

Figure 3.3: Epigenetics and Nutrition.

Source: https://blogs.biomedcentral.com/on-biology/wp-content/uploads/sites/5/2019/05/Figure-2-resized-300x210.png

Epigenetics and nutrigenomics are two fields of study that are becoming increasingly important in understanding how to maintain a person's overall health. It is possible to adjust epigenetic expression by methylating DNA or acetylating it, biotinylated or acetylating histones, or by doing all of these things at the same time. Alternatively, epigenetic expression can be modified by performing only one of these things at a time. Alternately, you could also perform both of these things at the same time, since this is very feasible.

The expression of a person's genes may be changed during a person's lifespan via the employment of epigenetic processes, and these changes can be handed down from one generation to the next. After guanosine, cytosine is the nucleotide that undergoes methylation in 5' CpG islands at a frequency that is the second-highest after that of guanosine. Guanosine is the nucleotide that undergoes methylation at a higher rate than any other. Both the transcription of genes and the structural integrity of the genome are jeopardized as a direct result of this. Methylation can take place in anywhere from sixty percent to ninety percent of the 5' CpG 3' islands that are present in mammals (Piroddi et al., 2017). Any alteration that occurs in the promoter regions of a gene, such as the ones mentioned above, will result in a change in the expression of the gene.

This is because the promoter regions are responsible for controlling how the gene is expressed. Because methylated 5′CpG 3′ islands attract capping proteins, which prevent transcription factors that would normally stimulate gene expression from reaching the gene, it is common to see an increase in methylation linked with gene silence or decreased gene expression. This is because capping proteins prevent transcription factors that would normally stimulate gene expression from reaching the gene. This is because capping proteins block transcription factors from accessing the gene, which would ordinarily increase gene expression. This is because the methylation process often focuses on methylated 5′CpG 3′ islands, which are a common target for the process.

When capping proteins are present, transcriptional factors, which are known to boost the quantity of gene expression, are blocked from accessing the gene. This keeps the gene from being expressed to its full potential. When this rule is violated, which happens when gene methylation affects the binding of inhibitory proteins, latent genes have the potential to become active once again. The methylation pattern of a gene is replicated, and then the pattern is passed on linearly from one copy of the gene to the next copy of the gene. Additionally, the capacity to methylate can be passed down via families (Pal, 2018). Because of this, the effects of methylation may not stop being felt in the body even after the process has been finished for an appreciable length of time. When DNA is firmly packed together, it is unable to access the genetic information that it contains because histone proteins prevents access to the areas of DNA that are essential for controlling gene expression.

This results in DNA being unable to read the information that it contains. The steps of acetylation and methylation, in addition to maybe biotinylation, need to be carried out for them to be uncoiled. Since this moment in history, transcription factors can enter gene promoters and activate the genes that they regulate. The methylation of 5' CpG 3' islands in gene DNA, which is the root source, is typically what causes these histone changes. This is because these islands are located within genes. Alterations in the availability of methyl groups in the diet have the potential to bring about long-term alterations in gene methylation, which, in turn, have the potential to influence gene expression and phenotype. It is quite likely that foods will influence DNA methylation because nutrients affect the availability and activity of the biochemical pathways that are involved in the process of DNA methylation. The metabolism of one carbon requires the participation of nutrients.

Some of the nutrients that could be on this list are choline, folate (which is a B vitamin), methionine, and vitamins B-12 and B-6. In addition to it, there is also a foliate that may be found here. It has been proven that depriving rats of dietary methyl (obtained via folate, choline, and methionine) results in broad global hypomethylation, as well as hepatic fat deposition, cirrhosis, and hepatocarcinogenesis happens even when there are no chemicals or medications in the rats that may cause cancer. Biological processes and disease-related phenotypes can be changed in response to environmental variables such as food since epigenetic modifications can be reversed (Neeha & Kinth, 2013). After all, one of the key mechanisms responsible for this phenomenon is epigenetic alteration.

3.4 GENETIC VARIATION

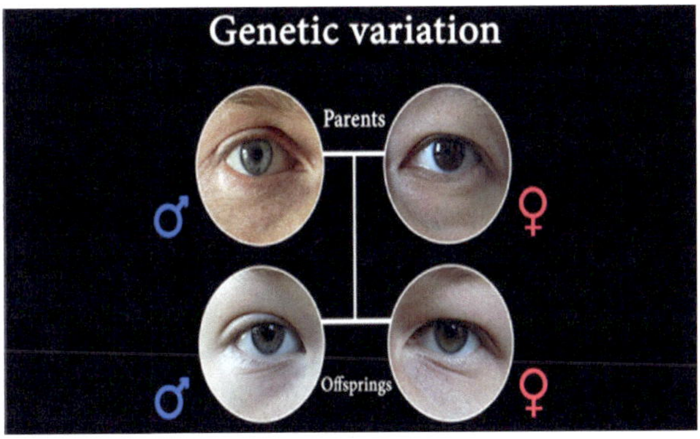

Figure 3.4: Genetic variation.

Source: https://www.proprofs.com/quiz-school/story.php?title=genetic-variation-test

There are over 50,000 single-nucleotide polymorphisms or SNPs in the human genome that are both homozygous and heterozygous. Some SNPs can alter the structure and function of proteins by altering the expression of genes or the products those genes produce. On the other hand, recent discoveries have made it increasingly plausible that one day we may be able to understand how millions of SNPs contribute to the dietary requirements of an individual. It is a matter that presents a great deal of difficulty. Despite the significant reduction in the price of genotyping, there is still a significant financial outlay involved in the process of genotyping all of the markers (McNamara, 2010). This is because there are so many markers that can now be obtained for each gene. As a consequence of this, researchers are required to select a set of genotyping markers to cut down on the number of SNPs that need to be investigated as part of their projects. Haplotype blocks are combinations of SNPs that have been found in a significant number of people who are not related to each other. Certain regions of a gene can be analyzed to uncover haplotype blocks, which are combinations of SNPs. Every single one of these SNPs is connected in some way to the others, and they are passed on from one generation to the next.

There have only been a few common SNPs that have been linked to differences in nutritional requirements. For foliate to be metabolized, an enzyme called 5, 10-methylenetetrahydrofolate reductase, more commonly abbreviated as 5-MTHF (MTHFR), must be present . Homozygous individuals have elevated levels of plasma homocysteine even if they get a sufficient amount of folate in their diet. The MTHFR gene's common SNP, the C677T allele, has been shown to reduce enzyme activity. This SNP is found in between 15 and 30 percent of the overall population. One more illustration of this would be the research on SNPs, which found that patients whose diets were deficient in choline had an increased risk of organ damage or failure (Mustapa et al., 2020). While some individuals were found to be free of fatty liver disease or other injuries to their liver or muscles, others were found to be affected by these conditions. A diet low in choline caused premenopausal women who carried a common genetic variant (MTHFD1-G1958A) to experience 15 times more symptoms of choline deficiency than the women who did not carry this variant. This was the situation even though the MTHFD1-G1958A SNP is a relatively uncommon variant.

In contrast, the risk of having a child born with a neural tube defect was increased by a factor of four among women who carried this SNP compared to the risk among women who consumed an adequate amount of choline during their pregnancies. It would be interesting to find out if the G1958A SNP in MTHFD1 is present in patients who are considered to be at high risk. In response to estrogen, the PEMT gene in the liver produces a protein that stimulates endogenous choline synthesis. This results in increased levels of endogenous choline. An SNP (single nucleotide polymorphism) in the PEMT gene promoter called rs12325817 is responsible for the increased risk of organ dysfunction in females brought on by choline deficiency. This risk factor was discovered as a result of research conducted on both genders. The overall population was found to contain 0.74 percent of this variant allele on average. The term "variation" is used to differentiate this common allele from the wild type, which was produced by a codon that is only ever found in mammals of a smaller size.

The wild type was formed by a codon that is only ever found in mammals of a smaller size. Research is also being done to investigate other genetic factors that play a role in determining how much choline a person needs to consume in their diet. Gene expression and the production of specific proteins like enzymes, hormones, and other biologically active molecules are both influenced by translation, post-transcriptional modifications, and changes in the quality of translation. As a consequence of these physiological processes, the production of enzymes, hormones, and various other biologically active proteins takes place (Afman & Müller, 2006). The production of a particular kind of protein in an animal's body is the culmination of the process known as gene expression, and that protein serves a distinct physiological or biochemical function. Gene expression can be affected by several different factors, including transcriptional, translational, posttranscriptional, and translational modification factors, as well as their regulators and environmental factors. Nutrients, which are environmental factors, have the potential to influence gene expression in several different ways, including the following:

3.5 THE EFFECT THAT CARBOHYDRATES HAVE ON THE EXPRESSION OF GENES

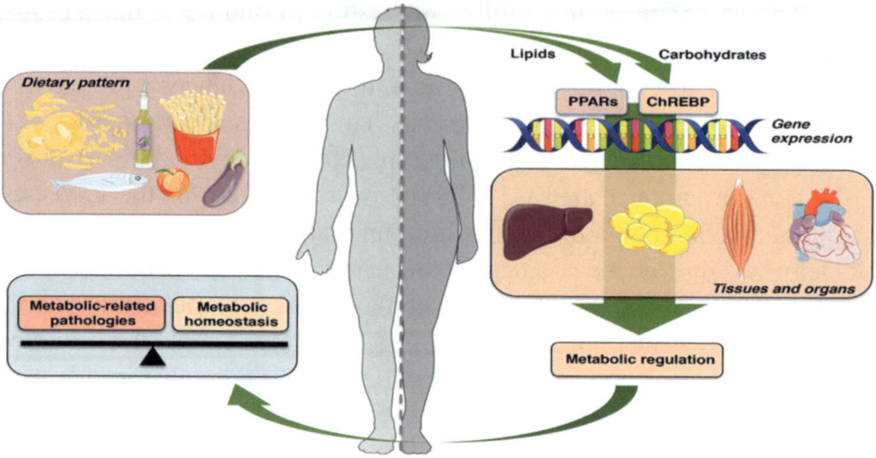

Figure 3.5: Effect of Carbohydrates on Expression of Genes.

Source: Bravo-Ruiz I, Medina MÁ, Martínez-Poveda B. From Food to Genes: Transcriptional Regulation of Metabolism by Lipids and Carbohydrates. Nutrients. 2021; 13(5):1513. https://doi.org/10.3390/nu13051513

When rats are given a diet high in calories starting at a young age, they have a greater chance of becoming obese and developing metabolic syndrome. It's possible that their bodies won't be able to adjust to the high number of calories contained in the meal. The levels of the oxygenic neuropeptides NPY and Ag RP in the arcuate nucleus of obese individuals were found to be significantly lower than in non-obese individuals. This was the case, even though the increase in total body weight was not nearly as significant.

A study titled "Hepatic Angiotensinogen Gene Expression in Hyperglycemia" found that hyperglycemia led to an increase in the level of AGT gene expression in the liver by a factor of two. Consuming a variety of carbohydrates, such as sucrose, fructose, galactose, and glycerol, can increase lactase activity as well as LPH (lactase/phlorizin hydrolase) mRNA levels in the jejunum (Masotti et al., 2010). These changes can take place in a short amount of time. These kinds of observations have been made on a variety of subjects, including humans and animals. The ingestion of a meal high in glucose led to an increase in lactase activity while simultaneously leading to a decrease in LPH mRNA concentrations.

The level of LPH mRNA in the cells was unaffected by the presence of A-methyl glucoside because this compound cannot be metabolized. In this experiment, rats that were given sucrose as a source of carbohydrate in their diet had higher levels of LPH mRNA than rats that were given methyl glucoside. One part of research concluded that fructose, one of the two monosaccharides that make up sucrose, has a greater potential than glucose to increase levels of LPH mRNA. This was the conclusion reached by the researchers. As a consequence of this, it is highly probable that fructose, which is a byproduct of the hydrolysis of sucrose, is directly responsible for the increases in LPH mRNA levels that are induced by sucrose. It is possible, given these data, that certain carbohydrates or the common metabolite(s) of these carbohydrates have a significant influence on the LPH gene.

3.6 GLUCOSE REGULATED GENE EXPRESSION

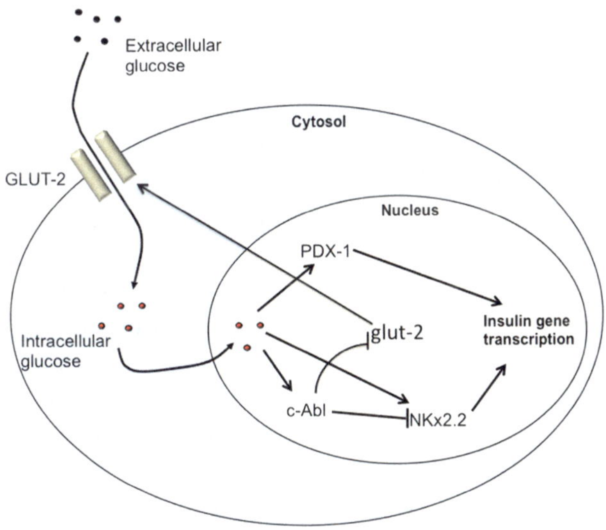

Figure 3.6: Glucose Regulated Gene Expression.

Source: Xia, Changqing & Zhang, Pengcheng & Li, Shiwu & Yuan, Lihui & Xia, Tina & Xie, Chao & Clare-Salzler, Michael. (2014). C-Abl Inhibitor Imatinib Enhances Insulin Production by β Cells: C-Abl Negatively Regulates Insulin Production via Interfering with the Expression of NKx2.2 and GLUT-2. PloS one. 9. e97694. 10.1371/journal.pone.0097694.

One of the most illustrative examples of how organisms have developed regulatory systems to cope with shifts in the availability of dietary resources is provided by glucose, which is the monosaccharide that is found in nature in the greatest abundance. Glucose is the monosaccharide that is found in nature in the greatest abundance. In yeast, the presence of glucose causes an increase in the expression of genes involved in metabolism, while at the same time causing a decrease in the expression of genes involved in the utilization of other carbon sources (Mathers, 2017). At this point, a comprehensive understanding of the process by which glucose possesses the capacity to impact gene expression in yeast has been achieved. It is also more difficult to respond to meal glucose because there are metabolic and hormonal impacts, such as pancreatic stimulation of insulin production and suppression of glucagon secretion in mammals.

Both of these things make it more difficult to respond to meal glucose. Because of these consequences, it will be harder for your body to respond to the glucose in your diet. Because of the simultaneous action of these two factors, this result has come about. Glucose is the primary physiological stimulus that governs the synthesis and release of insulin by beta cells in the pancreas, and the pancreas responds to glucose by exorcising insulin into the bloodstream. While glucose induces stimulation of L-type pyruvate kinase, acetyl-CoA carboxylase (ACC), and the fatty acid synthase enzyme, sugar causes the genes that code for the gluconeogenic pathway to be downregulated. L-type pyruvate kinase is an enzyme that helps convert pyruvate into fatty acids. Carbohydrates, in contrast to insulin and glucagon, were only recently found to have a part in transcriptional regulation, even though this is something that has been understood for quite some time. This is even though insulin and glucagon have been there for much longer. Researchers have found genome-binding complexes not only in the liver but also in beta cells. These genome-binding complexes play a crucial role in regulating gene expression in response to glucose.

3.7 GLUCOSE-6-PHOSPHATE DEHYDROGENASE

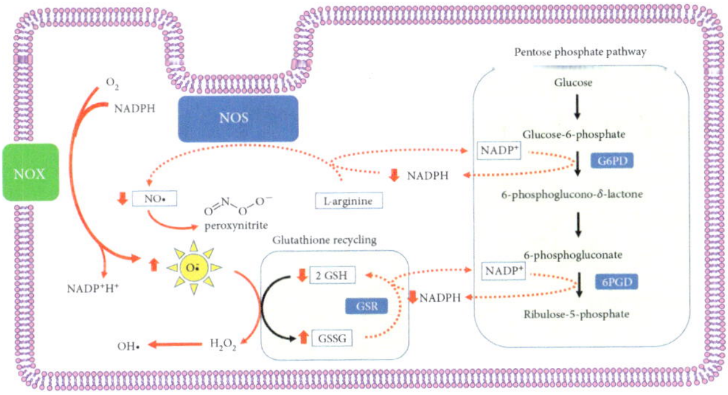

Figure 3.7: Glucose-6-Phosphate Dehydrogenase.

Source: Dore MP, Parodi G, Portoghese M, Pes GM. The Controversial Role of Glucose-6-Phosphate Dehydrogenase Deficiency on Cardiovascular Disease: A Narrative Review. Oxid Med Cell Longev. 2021 Apr 29;2021:5529256. doi: 10.1155/2021/5529256. PMID: 34007401; PMCID: PMC8110402.

Glucose-6-Phosphate Dehydrogenase is an enzyme that is a member of the lipogenesis enzyme family. It is also known by its enzyme abbreviation, G6PD. Investigation into how cells react to the presence of carbohydrate energy as opposed to triacylglycerol energy is one of its possible uses. Because it is engaged in a variety of distinct metabolic pathways, the enzyme that is commonly referred to as glucose-6-phosphate dehydrogenase, or G6PD for short, is a particularly remarkable example of its kind. These processes include the development of the cell, protection against oxidative stress, the generation of lipids, and the decrease of biosynthesis. G6PD activity was shown to be increased by glucose, whereas polyunsaturated fats were found to decrease G6PD activity. Glucose was found to enhance G6PD activity, while polyunsaturated fats decreased G6PD activity (Loor et al., 2015). These variations in G6PD activity are likely the result of interactions between the 331 genes and the G6PD nutrient. Such interactions have been demonstrated to occur. In contrast to the vast majority of lipogenic enzymes, whose gene transcription rates are known to be extremely sensitive to changes in dietary composition, the gene transcription rates of G6PD are known to be controlled by post-transcriptional mechanisms. This is in contrast to the fact that gene transcription rates of lipogenic enzymes are known to be controlled by pre-transcriptional mechanisms. During this stage of the process, the nucleus is responsible for producing mature mRNA; however, the efficiency with which the precursor G6PD transcript is spliced is also a key element in this phase of the process.

3.8 DIETARY FAT AND GENE EXPRESSION

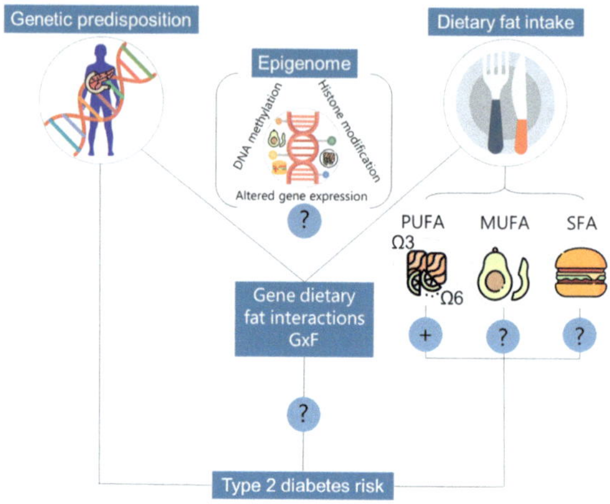

Figure 3.8: Dietary Fat and Gene Expression.

Source: https://www.researchgate.net/profile/German-Carrasquilla/publication/337014950/figure/fig1/AS:962677115469824@1606531518366/Interplay-between-genetic-predisposition-and-dietary-fat-intake-in-the-development-of.png

A diet must always include fat since it is a type of macronutrient that is necessary for normal growth and development in all forms of life and because this nutritional need is always present. It is common knowledge that the fat that is consumed in one's diet has a significant impact on the expression of genes, which in turn affects not only the processes that are involved in metabolism but also the rate at which cells develop and the total size of the body that an organism possesses (Banerjee et al., 2015).

This is an adaptive response that enables the quantity of fat ingested as well as the kind of fat to affect the level of gene expression. Researchers have discovered that the presence of fatty acids can influence the activity of certain transcription factors in bacterial cells, amphibian organisms, and human beings. In mammals, these proteins have been identified in the following four forms: peroxisome proliferator-activated receptors (PPARa, b, and g),

HNF4a, and SREBP1c (sterol regulatory element-binding protein). These proteins are denoted by names such as nuclear factor kappa B (NFkB) and sterol regulatory element-binding protein 1c (SREBP1c), for instance sterol regulatory element binding protein. Direct fatty acid or fatty acyl-coenzyme A binding, or oxidized fatty acid-binding; oxidized fatty acid regulation of cell surface receptors and activation of signaling cascades targeting the nucleus; or oxidized fat regulation of intracellular calcium levels, which influences cell signaling cascades targeting the nucleus. Direct fatty acid or fatty acyl-coenzyme: Direct binding between fatty acids or fatty acyls and coenzyme A- fatty acid that has been oxidized.

The rate at which cells metabolize fatty acids is influenced by a variety of distinct variables. These factors include: In this context, "factors" refers to both the quantity and the composition of the fat that is ingested, as well as the kinetics and competitive responses of oxidative pathways and membrane receptors that may be found in cells. It is also controlled by the amount of specific nuclear and membrane receptors that are present in cells, in addition to the engagement of specific transcription factors during the process of gene expression (Ladeira et al., 2018). This control is in addition to the fact that it is controlled by the engagement of specific transcription factors. In addition to regulating cytokines, adhesion molecules, and eicosanoids, these systems are also responsible for the metabolism of glucose and lipids. Additionally, these systems have an impact on the process of cell differentiation as well as cell development. The effect that the genome has on the metabolism of fatty acids has led to the discovery of new information regarding the role that dietary fat plays in both health and illness.

This information was discovered as a result of the discovery that the genome affects the metabolism of fatty acids. According to the findings of this study, PUFA, which is more generally known as polyunsaturated fatty acid, both represses and changes the post-transcriptional expression of SREBP-1. PUFA may also be referred to as polyunsaturated fatty acid. The consumption of PUFA brought modifications to the levels of SREBP-1 mRNA that were discovered in BeWo cells. The mechanism that is most likely responsible for PUFA's ability to inhibit gene transcription and reduce the amount of mature SREBP-1 protein is that they increase the amount of cholesterol regulatory pools that can be found within the cell. This is the mechanism that is responsible for PUFA's ability to inhibit gene transcription. When there is a bigger number of mature SREBP-1 proteins in the nucleus of the cell, the rate of enzymatic cleavage increases, which causes the cell to divide at a faster pace.

This can be accomplished by cultivating the cells on LPDS media that are free of cholesterol during the process of cell growth. Because there is no effect of n-3 fatty acid (FA) on leptin mRNA levels in the presence of growing mature SREBP-1, it would appear that the PUFA-induced effect on leptin expression may be attributable to a reduction in the quantity of immature SREBP-1. This is because there is no effect of PUFA on leptin expression. The fact that the PUFA-induced impact on leptin expression takes place demonstrates this point. It has also been proven that variations in the levels of plasma leptin in adults may be brought about by the consumption of foods high in fatty acids (Benítez et al., 2017). In addition, they showed that PUFA lowers the expression of the leptin gene in both *in vivo* and *in vitro* settings and that this influence is mediated through pathways associated with lower levels of PPARg and SREBP-1 gene expression. This was demonstrated by the fact that PUFA reduced the expression of the leptin gene. In both of the several experimental conditions, this was demonstrated to be the case.

3.9 VITAMINS AND GENE EXPRESSION

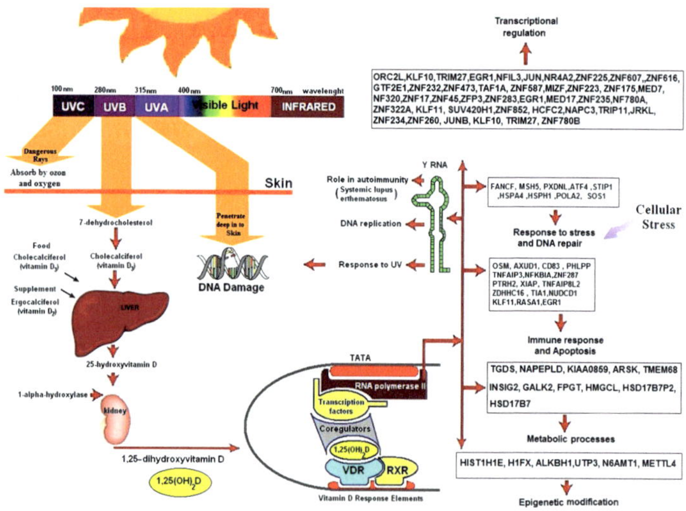

Figure 3.9: Vitamins and gene expression.

Source: Hossein-nezhad A, Spira A, Holick MF (2013) Influence of Vitamin D Status and Vitamin D_3 Supplementation on Genome Wide Expression of White Blood Cells: A Randomized Double-Blind Clinical Trial. PLoS ONE 8(3): e58725. https://doi.org/10.1371/journal.pone.0058725

Animals those do not produce their milk have a far lower requirement for the number of vitamins and other nutrients that they take in. Ruminants have the potential to generate some vitamins within their bodies if certain pre-vitamin components are present in sufficient concentrations. The majority of vitamins can influence how genes are expressed in the body. A lack of vitamins C and E, as well as oxidative DNA lesions and DNA lipid peroxide adducts, is associated with higher levels of DNA strand breaks, chromosomal breakage, and oxidative DNA lesions. Vitamin C and E deficiencies have been demonstrated to be associated with higher levels of DNA strand breaks. Other deficiencies, in addition to deficiencies in vitamin C and vitamin E, have been related to an increased risk of developing cardiovascular disease. Folate and vitamin B12 are necessary for the methylation of DNA, the production of dTMP from dUMP, and the recycling of folate (Kussmann & Affolter, 2009). In the absence of these nutrients, uracil is included, and an increase in the quantity of chromosomal breakage, and DNA hypomethylation will take place. Niacin requires a substrate, and that substrate is the enzyme known as poly (ADP-ribose) polymerase or PARP. PARP is involved in the repair and rejoining of DNA breaks as well as the maintenance of telomere length. It must be present for cells to become more prone to mutation as a result of exposure to mutagens and for there to be a rise in the frequency of DNA breaks that cannot be repaired. Vitamin A is required for the formation of both the growth factor insulin-like growth factor and the enzyme phosphoenol pyruvate kinase. Vitamin A is also required for normal vision. It is impossible to make these proteins without vitamin A. "Insulin-like growth factor" is the meaning of the abbreviation "IGF-1." Because it functions as a genetic cofactor, biotin is required for the creation of a wide variety of proteins that are of vital importance (enzymes). The expression of several genes in the liver is susceptible to being influenced by vitamin C's presence. The effects of vitamin A on PEPCK and biotin on the expression of the OTC (ornithine trans carboxylase) gene are known. Both of these genes are involved in the synthesis of ornithine trans carboxylase. Phosphoenolpyruvate carboxy kinase is an enzyme that is dependent on adequate quantities of vitamin A to function correctly. In the process of producing glucose, one of the most important steps in the gluconeogenic pathway is the conversion of oxaloacetate to phosphoenol pyruvate by PEPCK.

As a result of a lack of vitamin A, the RARE (retinoic acid-responsive element) chromosomal region will go through structural changes, which will make it more difficult for the co-regulators to connect and carry out their functions correctly. Because of RARE and the presence of the pre-initiation complex, RNA polymerase II has a diminished capacity to bind to the PEPCK promoter. This is because the pre-initiation complex is also present. It is possible to have an increase in gluconeogenesis if the levels of PEPCK in the body are decreased or abolished entirely (Bragazzi, 2013). Because of the propensity of this combination to activate the gene, the expression of the PEPCK gene is greatly increased in mice that do not have vitamin A, have low blood glucose, and have not eaten in a while. On the other hand, their ancestors who were deficient in vitamin A had extremely little or no expression of the PEPCK gene.

The quantity of vitamin A that is taken in can affect the expression of the IGF-I, IGF-IR, and IR genes that are found in a variety of organs. Inadequate amounts of vitamin A led to decreased levels of insulin-like growth factor I (IGF-I) in the blood and increased levels of IGF-I mRNA expression in the tissues, but it also led to increased levels of IGF-IR mRNA expression. Because of the significant decline in IGF-I mRNA levels and the concomitant fall in blood IGF-I concentrations, it is possible that animals who do not get enough vitamin A develop more slowly than usual.

3.10 HOW DNA IMPACTS FOOD DIET

Figure 3.10: Genes can influence food body response.

Source: https://qph.fs.quoracdn.net/main-qimg-48006247b2965eb5b8479d396ba58e35-lq

Your genetic make-up can affect whether or not you can tolerate certain meals, how quickly you become full, and the foods that you prefer to eat, all of which can lead to a distinct relationship with food. On the other hand, epigenetics, which is a fast-developing area of study, has demonstrated that external stimuli have the potential to alter the expression of our DNA. How big of an impact do our genes have on our physical well-being? The kinds of meals and drinks that are best for our bodies are heavily influenced by our genetic makeup. There is a genetic component to a person's flavor preferences, as well as their sensitivity to gluten and intolerance of lactose. According to experts, several genetic markers have been related to a range of behavior. These behaviors can be categorized as either positive or negative.

According to the findings of the current study, an individual's love for coffee may be the result of a genetic mutation in the gene that codes for bitter taste receptors in the human body. Although these individuals are more sensitive to bitter flavors, coffee continues to be one of their preferred beverages of choice. In addition to our genes, several other aspects of our environment play a role in determining how sweet or salty things taste to us. There is a possibility that some individuals have a genetic predisposition to consume meals that are heavy in sugar or salt. According to the findings of another study, a person's preference for oil over butter may be something that is handed down via their family. This may also alter the total number of calories, fat, and carbohydrates that you consume (Braicu et al., 2017). It has been shown via the discovery of genetic markers that individuals are genetically predisposed to eat in a particular manner, regardless of the cultural influences that exist or the accessibility of particular foods. This is the case although each of these things has an impact on the dietary choices that people make.

A person's weight may be influenced by their genes in two different ways: indirectly, via their diet, and directly, through their metabolism. Not only may your DNA affect the foods you choose to eat and the quantity that you consume, but it also has the potential to affect the creation of hormones and enzymes that are important for the metabolism of the body. When you use these strategies, your genetic make-up can affect how your body responds to nutrients, how likely you are to gain weight, and how quickly you burn calories. The fundamental chemicals that contribute to a feeling of fullness or hunger are determined by the genes that we inherit from our parents as well as the way those genes are expressed in our bodies.

It is theoretically feasible for these genes to be passed down from one generation to the next. The generation of the hormone ghrelin and the actions it exerts are both under the control of a single enzyme that was identified in 2008. Human ghrelin is the hormone responsible for causing people to feel hungry and compels them to consume more food. The elimination of leptin production in mice has been made possible thanks to the discovery of a genetic marker that the researchers had been looking for. The hormone leptin is responsible for communicating with the brain to let it know when a person is no longer hungry. This, in turn, affects how much food a person consumes. People who are obese tend to have higher levels of expression of the ob gene, which was discovered in humans.

For instance, DNA affects both the amount of food that is ingested and the path that nutrients take through the body. It has been discovered that gene expression plays a significant role in the process of muscle from fat differentiation. In addition, the body needs these components to keep the proportion of carbs, proteins, and fats in proper balance. Our physical selves are, to some extent, magnetized by the things that can fulfill their need. But what about the pressing needs that have only surfaced in the last little while? Is there a subset of the population that, because of their genetic make-up, is more likely to exhibit a certain character than others? There is no doubting the important part that genes play; this much is certain (Brennan & de Roos, 2021). According to the findings of one part of study, it has been hypothesized that differences in the activity of endorphins, which are molecules that are encoded in genes, may boost the desire to consume food. In another line of study, the neural circuitry of the brain was investigated, and genetic networks that govern food desires were found. The same circuits in the brain, which are governed by genetics, are responsible for impulsive behavior as well as compulsive behavior and addiction.

Your genetic make-up is one of the elements that can affect your height and weight, along with several other factors. Genes can change the composition of the body by affecting both metabolic and nutritional processes. The question of whether or not your DNA influences the food you consume is equally as essential as the food itself.

Environmental influences, such as a person's food and nutrition, have been shown to affect the way an organism's genes are expressed through the study of epigenetics. It would appear that people have an epigenetic inheritance, which means that the effects of the environment on genes may be passed down from one generation to the next. According to the findings of several research articles, the impacts are different for each individual by their age and the stage of development they are now in. The modifications may be undone, as demonstrated by the findings of an extra examination that was conducted.

According to the findings of several studies, our eating habits are among the most significant environmental elements that have the potential to affect how our DNA functions. One of the most powerful epigenetic processes is called DNA methylation, and it is precisely this process that determines how our DNA is expressed and how it functions in our bodies. In the past, methylation has been connected to both chemotherapies for the treatment of cancer as well as the process of aging. According to research, a person's diet may influence the way their DNA is methylated, which suggests that a person's diet might influence their risk of developing cancer (Kussmann & Van Bladeren, 2011). It has been shown that several vitamins and minerals are essential for the formation of specific DNA methylation patterns, which have been linked to an increased likelihood of getting sick.

These patterns can be influenced by a person's diet. Memory loss has been connected to choline shortage, for instance, as an example of a likely cause of the condition. Studies conducted on animals indicate that the relationship between nutrition and gene expression is significant even in the absence of human research on micronutrient intake and DNA methylation. This is the case although human research on these topics has not been conducted. The goal of scientists who investigate the epigenetic effects of food is to one day create dietary treatments that might assist people in maintaining their health in the future. The structure of DNA is likely more malleable than the majority of people realize; the fact that food has the power to change gene expression in a variety of different ways is evidence of this. The meals that a person consumes regularly might affect the individual's genetic composition.

CHAPTER 4

HISTORY OF NUTRIGENOMICS

CONTENTS

4.1 ORIGIN OF NUTRIGENOMICS

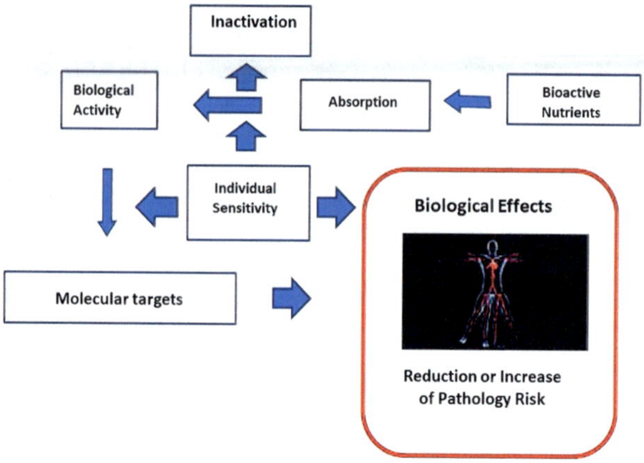

Figure 4.1: Nutrigenomics studies can reduce the risk of disease.

Source: https://www.oatext.com/nutrition-and-nutrigenomics-an-overview.php

The idea that diet affects health is an old one. Nutrigenomics integrates known collaborations between food and acquired genes, known as 'innate digestion errors,' often treated by diet manipulation. One such model is Phenylketonuria (PKU); brought about by a variation (mutation) in one gene. Affected people should keep away from food with amino acid phenylalanine (Kang, 2012). Another illustration is lactose intolerance, most grown-ups on the planet are lactose intolerant, implying that they can't process milk products, since the gene encrypting lactase, the enzyme that disintegrates lactose, is typically 'switched off after weaning." Nevertheless, around 10,000 to 12,000 years ago, polymorphism in one DNA nucleotide emerged among northern Europeans.

The single nucleotide polymorphism, or SNP, resulted in continuous lactase gene expression into adulthood. This was beneficial since individuals with SNP may use nutritious dairy items in areas with brief growing seasons, and with the transformation in sub-atomic genetics during the late 20th century, researchers set out on a mission to recognize different genes that interrelate with dietary elements. By the 1980s organizations were commercializing nutrigenomics. The 1990s Human Genome Project, which sequenced the whole human genome DNA, kicked off the study of nutrigenomics. By 2007, researchers were discovering various interrelations between nutrition, genes, and disease.

Nutrigenomics introduces new terminology, innovative experimental strategies and a fundamentally new strategy to nutrition research, for example, high-throughput technologies which allow for the universal research of gene expression in cells and organisms. Nutrigenomics needs cooperation from specialists in genetics including the sectors of food science, public health and culinary. It's rather simple to make great-tasting food (Irimie et al., 2019). Add some margarine or lard inside, and the taste improves. The test is on how to remove the fat and make refreshing yet great tasting food. Thus, a change in public health is significantly required, and with a rising frequency of obesity and terminal illnesses like type II diabetes, nutrigenomics could become the panacea later on.

4.2 BEGINNINGS OF MODERN NUTRIGENOMICS

The word "nutrigenomics" was first defined in 2001 by Pelegrin (2001) and afterwards, showed up in a 2002 study by Van Ommen and Stierum (2002). Starting around 2001 until the present, a PubMed database reveals 'nutrigenomics' as a search word with roughly 1005 articles. This single search shows a rapid way to the nutrigenomics future and the fields which must be studied further. Adapted nutrition shall be the future concerning development and endorsing an eating regimen for people depending on their genome and their genetic qualities. As per a research article on the Greek public population (N = 1504) and health specialist (N = 87) samples, it was discovered that there is more requirement for learning the emergent nutrigenomics science (Pavlidis et al., 2012). During the review, just 11.5% of the respondents were encouraged to embrace a nutrigenomics test, suggesting that there is certainly not a widespread application among nutritionists, physicians or health specialists. Furthermore, 23.5% of the subjects were constantly requesting their healthcare suppliers for nutrigenomics tests, whereas the higher 80.5% of physicians and nutritionists were ready to suggest a nutrigenomics analysis (Pavlidis et al., 2012). Illnesses, for example, diabetes, obesity, high cholesterol and fatty lipid levels were considered related to the genetic profile of a person as captured by the opinions of health experts and the general population. Specifically, 76% of the general population mentioned that the use of a custom diet developed by their genetic makeup would be valuable (Pavlidis et al., 2012).

Nutrigenomics studies are still continuous. A present metanalysis on 38 nutrigenomics genes revealed that there is no clear relationship between the genes typically analyzed in nutrigenomics tests, and various diet-related diseases, though in some situations there are few studies relating them (Pavlidis et al.). A relationship between APOA5 c.- 1131C>T and fatty acids and APOA-V levels was mentioned, when the relative impact of legumes, whole grains and refined rice was explored in recently diagnosed patients with diabetes type 2 (Kang et al., 2014). Around the same time, Li et al. (2014) explored the impact of different polymorphisms (rs662799, rs3135506, rs2075291, rs2266788) of the APOA5 genes on fatty acid levels in 1174 Uyghur (combination of East Asian and Caucasians) subjects. While a dysregulation of fatty acid levels was obvious, the requirement for additional exploration likewise became clear. Furthermore, a comparable report, showed an age-related relationship in mice among fatty acids and the APOA5 c.- 1131T>C polymorphism, however for those with the TT allele and not the CT or the CC alleles (Kim et al., 2014). These models show how complex nutrigenomics research is as far as impacts, genetic material and inclination in different populations and environmental elements that could impact their performance and quality is concerned. A quality dietary routine connection has been found, when two of the most known genes (the FTO and the MC4R) were concentrated on reviewing the adherence to the Mediterranean Diet for two polymorphisms (rs9939609 and rs17782313 separately), despite no relationship to diabetes type 2 found (Ortega-Azorín et al., 2012). The body weight in youngsters with diabetes could be affected by the presence of the rs9939609 allele (Luczynski et al., 2014).

4.3 EARLY FOUNDATIONS

The trailblazers of nutrition research defined the energy ratio of food and furthermore assisted with upsetting misconceptions about different infections that plagued humanity.

Nutrigenomics — and the remainder of modern nutrition science — relies on the foundations established during the late 18th century.

This isn't imply that nobody had considered beforehand how food functions. The early civilizations of Rome, Persia, China, Egypt, Greece and India knew about the association between food and wellbeing. They all followed their unique food rules, large numbers of which are as yet relevant today, for instance, early Egyptians used garlic restoratively. Some modern food-insights trace back centuries. The ancient Greek doctor Hippocrates suggested that food must be well chewed prior to consumption, and taken moderately to maintain great health (Bionaz et al., 2015).

However, before the 18th century, there was minimal scientific research concerning the structure of food, including how the body metabolizes it. The early researchers were "reliant on experimental opinion". Their technique was 'feed and observe'. However, it was French scientist Antoine Lavoisier, considered the pioneer of modern chemistry, who initially performed the study that prompted modern-day nutrigenomics.

4.4 FOOD AS FUEL

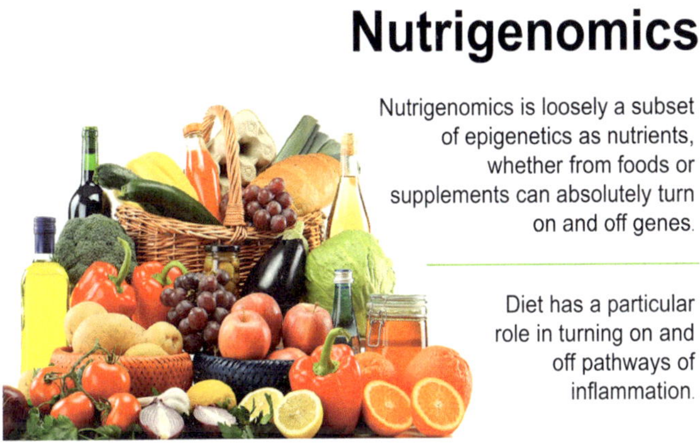

Figure 4.2: Good diet is effective in maintaining a healthy body.

Source: https://www.tandfonline.com/doi/pdf/10.1080/07315724.2019.1582980

Lavoisier was quite possibly the earliest researcher to develop lab equipment to test what becomes of food after consumption. Before his project, researchers recognized that the heaviness of ingested food surpassed that of urine and feces. They credited this loss to sweat. However, Lavoisier accepted that food was energy and that the human body, similar to the fuel-consuming motors being created at that age, must release carbon dioxide as a result of combustion. He believed that exhaled carbon dioxide represented this lost content.

To test his hypothesis, during the mid-1780s Lavoisier designed a new form of an instrument— the ice calorimeter. It was made out of an external shell loaded with ice, to keep a steady temperature of 0 °C, enclosing a chamber encasing a guinea pig. The creature's body heat dissolved the ice. Through weighing water streaming out of the calorimeter, Lavoisier had the option to gauge metabolic temperature and contrast it with the heat delivered by a lit candle or fired coal.

His hypothesis proved right. Lavoisier proclaimed: "respiratory gas exchange resembles a combustion such as a candle combustion."

In the modern calorie-conscious world, this doesn't seem like a very remarkable disclosure. However, during that time it was a leap forward. "It was hypothetically critical to understand that the body required energy to work and that one significant role of food is actually supplying it," mentions Elizabeth Neswald, a science history scholar at Ontario, Canada's Brock University. "It was a reason for figuring out what somebody needs for survival; what causes weight gain, what permits physical labour, and the connection between food and physical labour."

Lavoisier's research further underlined the significance of food constitution and understanding that excrement, urine, sweat and respiration are fundamental aspects of the equation.

Early nutrition researchers spent most of their time examining others' excrement.

"These early nutrition researchers spent most of their time — or their collaborators' time — investigating and examining others' stool," claims Neswald. "In nutrition tests, it was fundamental to evaluate the variations between input and output — food entering and other products released."

This technique referred as 'balance trials', was spearheaded during the 1830s by French scientist Jean-Baptiste Boussingault. He led balance tests for nitrogen — a constituent component of proteins — by contrasting the nitrogen ratio of oats, hay and potatoes given to horses and cows with the creatures' stool and, for cows their milk content (Castro et al., 2006). He demonstrated that animal feed comprised adequate nitrogen to fulfill bodily needs, completing speculation that extra nitrogen was assimilated from the environment.

4.5 MACRONUTRIENT RESEARCH

During the mid-nineteenth century, researchers discovered that the primary components of food are carbon, hydrogen, nitrogen, and oxygen, and separated food components into four primary varieties: fats, protein, carbohydrates and water. However, the chemical constitution of the first three categories was obscure.

Justus von Liebig, a German chemist, took the next jump forward. The bright von Liebig (chosen University of Giessen professor at 21yrs) designed the 'kaliapparat', unique glassware for carbon analysis in organic matter.

Von Liebig's research center, apparently the first research laboratory, attracted researchers from across the globe. He helped coach an age of nutritional specialists whose contributions would persist into the mid-20th century. During the 1860s, for instance, two of von Liebig's apprentices —chemist Max Joseph von Pettenkofer and physiologist Carl von Voit— got financing from the Bavarian state to develop a cutting-edge breathing chamber sufficiently large to hold an individual. The chamber would quantify the daily equilibriums of carbon and nitrogen and thus determine human protein needs.

Neswald mentions that the vast majority of nutrition research during this period concentrated on people's health, though later on discovered the most affordable, simplest ways to feed the "impoverished populations" to forestall the food crisis. Von Voit, mentions that Neswald, visited prison facilities and workhouses "to evaluate what individuals were fed and what their health condition was, intent on offering nutritional guidelines".

The idea of food as nourishment, which contains significant dietary constituents, was additionally refined in the U.S. Agricultural scientist Wilbur Olin Atwater invested energy in von Voit's research facility as a postdoc, coming back to the U.S. in 1871 to lead nutrition science. Atwater took five years during the 1890s to develop a respiratory calorimeter bigger than von Voit's and capable of holding people for longer than one day (Hurlimann et al., 2014). His estimations were so accurate that his energy measure for protein, fat and starch are still utilized today. Atwater was the first to use the term 'calorie' as a food energy unit (A food energy calorie is 1000 thermal energy calories).

4.6 INCREASINGLY SMALLER SIZES

Researchers before long started to understand that apart from providing energy and macronutrients, food likewise played a rather delicate function in health and illness. Japanese doctor Takaki Kanehiro, who studied during the 1870s at London's St Thomas' Hospital Medical School, was a rare exemption from the nineteenth-century German predominance of nutrition. "He became the first individual to demonstrate that beriberi emerges from malnutrition," remarks Katsuhiko Yokoi, a nutritionist at Japan's Seitoku University. Formally, beri beri was considered an infectious condition.

During the mid-20th century, different researchers across the world started to investigate relations between nutritional deficiencies and different diseases, including scurvy and rickets. Incapable of explaining these afflictions in regards to protein, fats, or sugars, a few researchers started suspecting the existence of a new class of food components.

In 1912, the Polish organic chemist Casimir Funk, while researching beriberi, isolated thiamine, the compound that protects from infection. He referred to the substance as a 'vital amine', later shortened to 'vitamin'.

The fight against scurvy is an illustration of research later refining a nutrition-based disease association. During the eighteenth, Scottish maritime doctor James Lind discovered that scurvy is preventable by consuming citrus fruits. However, he inaccurately believed that ocean air was to be faulted for the illness. Other mistaken ideas followed: in 1846, for instance, Scottish toxicologist Christison, Robert theorized that scurvy was brought about by lack of protein. Scurvy kept on being a sporadic issue well into the mid-20th century (Hesketh, 2013). It was only after 1932 that American biochemist Charles Glen King demonstrated that scurvy was brought about by a lack of L-ascorbic acid.

Animal research highlighted additional vitamin and illness-related discoveries. American biochemist Elmer Verner McCollum learned the German language to read the content of former nutrition specialists, which motivated him to investigate rodents. McCollum, from the University of Wisconsin, where he previously worked, researched principles around using cows as test models. Yet, McCollum persuaded his bosses to allow him to use smaller creatures. He purchased 12 albino rats from a rodent store and formed the first rat colony for lab experiments in the United States. In 1913, his research on these rodents made him classify the first fat-solvent nutrient, vitamin A, and later demonstrated that vitamin D, instead of vitamin A as some suspected, prevented rickets.

Demonstrating the connection between micronutrients and sickness didn't come without any challenges. US Public Health Service and disease control specialist Joseph Goldberger mentioned that pellagra, at that point a significant sickness causing diarrhea, dermatitis, dementia and demise, was diet-related and not, as traditional clinical assessment held at that point, an irresistible infection. In 1916, to make his statement, Goldberger and his colleague exposed themselves to a number of tests — they infused each other with blood from a pellagra victim, cleaned out the discharges of a pellagra patient's nose and throat and transferred them inward, and consumed capsules containing scabs of pellagra victims' rashes. But regardless of such intense exposure, they didn't cause pellagra (Gonzalez et al., 2015). Nonetheless, Goldberger couldn't locate the diet-based cause. It was an additional twenty years before U.S. biochemist Conrad Elvehjem discovered that pellagra was brought about by a lack of niacin (vitamin B3).

Countless micronutrients had been found by 1944 that some accepted nutrition had been wholly defined with nothing else to find. Yet, whereas the constituent food ingredients could have been coaxed out, their body effect was simply beginning to be appreciated.

From Lavoisier to von Liebig, to modern researchers like Jose Ordovas nutrition scientists have increasingly focused on smaller components. As researchers have examined further into biochemical processes of metabolism and function — opening secrets as they progress — they have likewise triggered new inquiries, until we reach the space of 'how our genes mix with the food we consume.' This is the issue scientists are still attempting to answer today.

4.7 FOUNDATIONS OF GENE-NUTRIENT INTERACTIONS

The human genome, including the genetic variation available within human populations, is to a limited extent a result of adaptive evolution to a frequently scarce and unconventional food supply. Single nucleotide polymorphisms or SNPs, which are regular, single base-pair variations in DNA sequence, constitute a primary type of human genetic assortment. Of the roughly 10 million SNPs found in the human genome, a good number are believed to have practical outcomes (e.g., modify the function/capacity of the protein item). SNPs emerge from the sequential procedure of DNA mutation and ensuing extension of the mutation inside a population (Guéant et al., 2014). Food nutrient exposures impact both of these processes. For instance, B-vitamin deficiencies hinder DNA metabolism/stability and increase DNA mutation rates (microorganism line and somatic DNA transformations) as do abundances of prooxidants, comprising iron. Moreover, the nutritional environment is capable of speeding up the development of fortuitous microorganism line DNA transformations inside a given populace, with the end goal that they amass and add to human genetic variation.

Without a doubt, different SNPs that impact nutrient use display genomic "markers" of such positive determination. For instance, a SNP situated close to the gene that encodes lactase allows SNP carriers to develop this enzyme throughout adulthood and hence continue tolerating milk. This SNP entered populations whose ancestors originated from areas animals could be nurtured safely and economically, like in Europe. Notwithstanding, different gene variants that emerged from positive selection are contemporary candidates for infection alleles; gene variations that allow adaptation to a particular environment may be malicious when the environmental conditions change, for example, the nature and quantity of the food supply. For instance, the HFE gene variant related to the risk for hemochromatosis might have benefitted iron-poor areas however gives risk for iron over-abundance in iron-rich conditions (Fenech et al., 2011). SNP developments likewise require that the related changes in biochemistry allow embryonic endurance in the intrauterine condition. Both malnutrition and various gene variations that impair nutrient digestion and usage are risk factors for early fetal termination.

Nutrients and related bioactive food constituents can likewise monitor gene expression. All creatures have gained the capacity to sense and adjust to their nutrient environment, through changing the expression of the proteins that work in metabolic and signal pathways. Outstanding models include, however, are not restricted to, the moderation of vitamin A and vitamin D gene transcription through association with their separate atomic receptors. This capacity of nutrients to interrelate with the genome is a fundamental aspect of organismal evolution. Nutrients may evoke transient adjustments in gene expression and impact more durable and holistic genome restructuring incidents that can be acquired (that is, passed over generations). The phrase epigenetics means the inheritance of attributes through systems that are autonomous of DNA basic sequencing and consist of the gene expression designs or expression levels that add to phenotypic disparities in people, including monozygotic doubles.

The embryo particularly seems to be susceptible to nutrient-based gene expression adaptations, a phenomenon known as metabolic programming or imprinting. These adaptations happen inside critical windows during embryonic development and can continue into adulthood. The related changes in digestion resulting from these reprogramming factors are believed to allow for endurance in the uterus in poor nutrient conditions, however, predispose the person to metabolic sickness in adulthood. This maternal nourishment process, fetal epigenetic programming, adult-onset terminal sickness is the premise of the fetal beginnings of adult infection hypothesis, which recommends that good nutrition matters early in life to prevent the risk of negative health results in adult life. This hypothesis, which was initially upheld only by epidemiological relations, has now been approved in animal studies (Ferguson et al., 2010). These studies show that early nutrition exposures expanded the risk in adult life for obesity, hypertension, and insulin management, which are the predecessors of adult terminal diseases like cardiovascular sickness and diabetes. The genome, thusly, can moderate diet. Genetic variation in epigenetic programming may affect nutrient absorption and usage (for example, hemochromatosis) and accordingly present differences in food/nutrient tolerances (for example, iron) and may add to the variation in human nutrient needs.

4.8 HISTORY OF MODERN NUTRITION SCIENCE

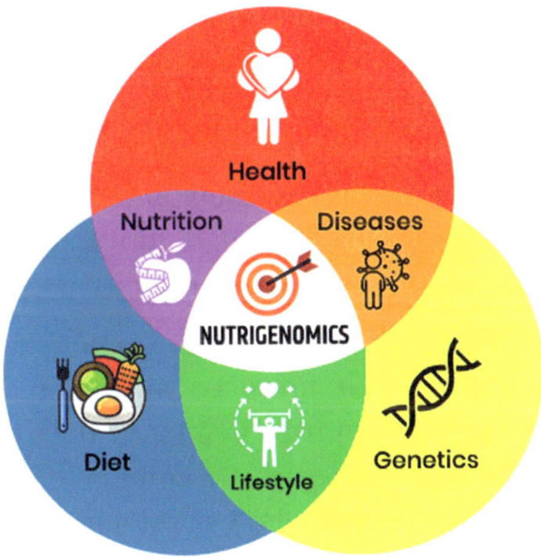

Figure 4.3: Nutrigenomics affects every aspect of our life, including health and diet.

Source: https://goqii.com/blog/nutrigenomics-you/

While food and nutrition have been researched for centuries, contemporary nutritional science is amazingly young. The original vitamin was separated and chemically described in 1926, around 100 years ago, introducing 50 years of revelation focused on a few nutrient deficiency conditions. Studies on the function of nutrition in complex non-transmittable terminal diseases, like cardiovascular disorder, diabetes, obesity, and tumors are more recent, advancing over the last few decades and particularly after 2000.

Historical records of nutrition science have been cataloged, concentrating on dietary guidelines, broad scientific developments, or specific nutritional treatments. Carl Sagan mentioned "You must understand the past to figure out the present;" whereas Martin Luther King, Jr stated, "We are not history makers. We are fashioned by history." This section defines key historical events in contemporary nutrition science that constitute the premise of modern comprehension of diet and health, explaining contemporary needs, new fads, and debates in nutrition science and strategy (Ferguson, 2016).

4.8.1 1910s to 1950s: Period of Nutrient Revelation

The first half of the twentieth century saw the grouping and synthesis of different fundamental vitamins and minerals and their application to forestall and treat nutritional deficiency-based illnesses such as beriberi, pellagra, scurvy, rickets, nutritional anemias and xerophthalmia. In 1913, Casimir Funk developed the concept of food "vital amine", beginning from the opinion that the mass of whole rice protected chickens from beriberi-like disorder. The "vital amine" or vitamin was originally isolated in 1926 and called thiamine, and thusly blended in 1936 as vitamin B1. In 1932, L-ascorbic acid (vitamin c) was isolated and authoritatively recorded, interestingly, to protect from scurvy, around 200 years after ship physician, James Lind, tested lemons for managing scurvy in mariners.

During the mid-twentieth century, all primary vitamins had been separated and synthesized. Their presence in human and animal studies demonstrated the nutritional premise of intense deficiency disorders and originally prompted dietary strategies to combat beriberi (vitamin B1), scurvy (vitamin C), pellagra (vitamin B3), pernicious anemia (vitamin B12), rickets (vitamin D), and related deficiency conditions. Notwithstanding, the chemical vitamin synthesis rapidly prompted food-based policies being supplemented by treatment using individual vitamin supplements (Fenech, 2008). The use and promotion of individual and packaged multivitamins to create solutions for deficiency set off a whole vitamin pills industry.

This new research on individual nutrient deficiency diseases likewise prompted the fortification of chosen staple food sources with micronutrients, like iodine in salt and vitamin B3 (niacin) and iron in bread and wheat flour. These strategies ended up being powerful at lessening the prevalence of different common diseases, such as xerophthalmia (vitamin A), goiter (iodine), anemia (iron) and rickets (vitamin D). Different foods all over the planet have since been added with, iron, calcium, phosphorus and essential vitamins (A, B, C, D), based on the synthesis of local staple food sources.

Being one of the incredible discoveries of nutrition history, this new science coincidentally emerged during the Great Depression and World War II, a period of boundless fear of food deficiencies. This prompted a further emphasis on avoidance of deficiency syndromes. For instance, the original recommended dietary allowances (RDAs) were an immediate result of these concerns, when the U.K. Medical Association, League of Nations and the US government independently commissioned researchers to create new minimum dietary needs necessary for World War II. In 1941, these original RDAs were declared at the National Defense Nutrition Conference, giving new rules for overall caloric and chosen nutrients such as phosphorus, iron, protein, calcium and certain vitamins (Fallaize et al., 2013). These historical occurrences established a reference point for nutrition science and policy suggestions to concentrate on a given nutrient related to particular disease conditions.

4.8.2 1950s to 1970s: Sugar versus Fat and the Protein Gap

Over the following 20 to 30 years, calorie malnutrition and particular vitamin deficiencies dropped considerably in top-income nations due to economic development and massive increases in minimal cost processing of staple food varieties fortified with vitamins and minerals. Simultaneously, the rising risks of diet based non-transmittable ailments started to be recognized, prompting new research guidelines. The focus was on two areas: sugar and dietary fats.

Early environmental studies and minor, short-term interventions, most noticeably by Frederick Stare, Mark Hegsted, and Ancel Keys, added to the far-reaching conviction that fat was a significant cause of coronary illness. Simultaneously, efforts by John Yudkin and others mentioned excessive sugar as the cause of coronary disease, hypertriglyceridemia, malignant tumor, and dental caries. Eventually, the focus on fat won scientific and policy acknowledgement, epitomized in the 1977 US Senate council report Dietary Goals for the United States, which suggested low fat, low cholesterol foods for all (Elsamanoudy et al., 2016). This was not without contention: in 1980, the US National Academy of Sciences Food and Nutrition Board reviewed the information and reasoned there was deficient proof to restrict total fat, dietary cholesterol and saturated fats across the population.

Some decipher these debates as proof of industry impact, and others as natural conflict and advancement of early science. More significant is that both dietary fats and sugar models depended on a nutritional prototype created to address deficiency conditions: recognize and seclude the single relatable nutrient, review its physiological impact, and measure its ideal intake amount to prevent infection. However, future research proved that such reductionist models are poor in interpreting non-transmittable diseases.

In less well-off nations, the fundamental objectives of nutrition strategy and proposals during this era remained intensifying calories and chosen micronutrients. In numerous ways, food varieties became seen as a delivery car for basic nutrients and calories. Appropriately, agrarian science and technology stressed production of minimal expense, shelf stable, and energy packed starchy staples like rice, corn, and wheat, with related nurturing and handling to separate and cleanse the starch maximally (Dimitrov et al., 2016). As in rich countries, these endeavors were joined by strengthening of staple food and food assistance projects to encourage the survival and development of babies and small kids in poor populations.

Researchers concentrated on malnutrition differing on the relative function of overall calories and protein in children and infant sicknesses like kwashiorkor and marasmus — additionally named "the protein-calorie deficiency illnesses." Sustenance for the "protein gap" model prompted the broad modern improvement of protein-enhanced formulas and correlative food sources for developing nations. Different researchers supported the fundamental job of calorie inadequacy and accepted that protein-enriched recipes and food sources shouldn't exchange breast milk. As one researcher wrote in 1966, "dollars and long periods of effort… into making these [high protein] food varieties would have been well spent on endeavors to protect the act of breastfeeding... being deserted elsewhere."

The discussion fundamentally ended when in 1975 key researchers in London and the US independently concluded from research evidence that an absence of food was the primary issue: "The idea of a global protein gap… is as of now not valid… the issue is fundamentally one of volume as opposed to nature of food."

This conclusion impacted subsequent endeavors to handle malnutrition in non-industrial nations. For instance, a formal UK advisory board of trustees on global nutrition aid suggested that "the main attack on malnutrition must be through the mitigation of poverty and must be channeled to projects that will create income among poor people, even where such programs meaningfully affect the public income of the nation concerned."

Nonetheless, the prior years of vulnerability had encouraged a global industry that kept on promoting formula and toddler food sources in low-income nations in view of their protein ratio and nutrition fortification. Moreover, nutrient supplementation techniques remained practical in forestalling or handling endemic deficiency conditions (Dimitrov, 2011). Hence, in spite of the change in scientific reasoning to concentrate on economic development, the significant emphasis remained or even progressed on offering adequate calories, most frequently as starchy staples, apart from vitamin supplementation and fortification.

4.8.3 1970s to 1990s: Diet Based Terminal Conditions and Supplementation

Speeding up economic development and transformation of agrarian, food handling, and food formulation methods kept on decreasing single nutrient deficiency conditions worldwide. Coronary mortality likewise started to fall in rich nations, however numerous other diet-related terminal illnesses were expanding, including type 2 diabetes, obesity and a few cancer types.

Accordingly, nutrition science and strategy rules in developed countries moved to attempt to manage terminal illnesses. Expanding on the 1977 Senate records, the 1980 Dietary Handbook for Americans was among the earliest such countrywide guidelines. The majority of the accessible data were gotten from less robust kinds of evidence, for example, from raw environmental examinations and economic tests utilizing substitute results, generally in healthy middle-aged men. More significantly, these examinations followed the deficiency condition model, to a great extent taking into account isolated single nutrients (Carlberg, 2019). As needs are, the 1980 dietary rulebook remained highly nutrient centered: "avoiding excess fat, saturated fat, and cholesterol; eat food varieties with sufficient starch and fiber; avoid excess sugar; stay avoid excess sodium. Global rulebooks were likewise nutrient focused. This prompted a multiplication of commercially viable food items low in fat, cholesterol, saturated fats and sustained with micronutrients, and the development of other nutrient-centered advances to decrease saturated fat like limited hydrogenation of vegetable oils.

Then, the international community focused on activities to reduce hunger and micronutrient deficiency in low-income countries. Significant micronutrient targets during this duration were vitamin A, iodine and iron. Evidence was growing that vitamin A supplements could prevent death from infection, like measles, and also prevent night visual impairment and xerophthalmia. Field trials offered a premise for WHO suggestions for extensive micronutrient supplementation, particularly during pregnancy, with iron and vitamin A, and for salt fortification with iodine to prevent goiter and genetic conditions like genetic hypothyroidism and hearing loss.

In view of these needs, the United Nations, national governments, including other global groups embraced portfolios for avoiding micronutrient deficiencies with supplementation and fortification, as well as integration of the developing evidence (Dall'Asta et al., 2022). Scientific research further centered around other environmental factors which might interrelate with micronutrients and dietary protein, for example, disease and related low sanitation, prompting ideas, for example, subclinical enteritis and malabsorption referred as "tropical enteritis," then "environmental enteropathy," and presently "environmental enteric disorder." Hence, in both low and high-income countries, for partially overlapping reasons, a nutrient-centered focus formed on scientific inquiry and strategy interventions.

4.8.4 1990s to Modern Times: Evidence Discussions, Diet Policies, the Double Burden

Among the most scientific developments of recent years was the emergence and completion of diverse, correlative, nutrition studies, including imminent observational cohorts, randomized clinical trials, and, as of late, genetic consortiums. Cohort research offered, interestingly, personal level, multivariable modified findings on a variety of nutrients, food types, and diet designs and a variety of health outcomes. Clinical preliminaries permitted further testing of open inquiries in designated, frequently high-risk populations, specifically impacts of separate nutrient enhancements and, all more lately, unique dieting plans. Genetic consortiums gave significant proof of genetic impacts on dietary decisions, quality eating routine communications influencing disease risk factors, and Mendelian randomization investigations of secondary impacts of nutritional biomarkers (Dang et al., 2014).

These developments were not without argument, specifically the general dissonance of discoveries between related studies and supplement preliminaries for certain vitamins on cardiovascular and cancers. Some specialists interpreted the conflict as proof of irredeemable deficiencies of observational research. Others mentioned it showed the risks of single nutrient methods to handle terminal diseases and mirroring the different systemic plans, with trials frequently centered on brief, supraphysiological dosages of vitamin supplements in risky patients, while observational analyses frequently centered on constant vitamin intake from food within the communities. As opposed to single nutrients, physiological mediation tests, cohort research, and randomized clinical tests offered more reliable proof for diet models, like low fat dietary plans (few key impacts) or Mediterranean and comparative food-based models (steady benefits). This concordance was upheld by progress in research techniques and improved comprehension of the reciprocal genes of various study designs.

Overall, these advances recommended that single nutrient hypotheses were insufficient in making sense of many impacts of diet on non-transferable diseases. This passed the the RDA system and other nutrient measurements intended to recognize the threshold for nutrient deficiency diseases, and toward complex environmental impacts of food sources and dietary patterns. These factors were progressively seen to reproduce joint contributions and relations between starch quality (e.g., fiber content, glycaemic index), fatty lipid profiles, protein forms, micronutrients, phytochemicals, food model, preparation and handling strategies, and additives (Chen & Zhao, 2013).

Strategic cohorts and dietary intervention tests showed that emphasis on total fat, a pillar of dietary rules since 1980, produced minimal quantifiable health advantages; on the other hand, nutritional suggestions for certain food varieties like red meats, eggs, and dairy products (example, based on dietary cholesterol, calcium and saturated fats) gave a false depiction of the noticed relations of these food varieties with health benefits. For weight reduction and glycaemic management, years of concentration on low-fat dietary regimens were addressed by the consequences of progress of imminent cohort studies, metabolic feeding research, and randomized tests, which demonstrated that food high in healthy fats delivered benefits, while foods high in starch and sugar created harm. This development was extended to the acknowledgement of the relevance of dietary patterns, for example, the traditional Mediterranean and vegetarian diets emphasise on fewer calories that stressed low processed food varieties like fruit products, nuts, beans, vegetables whole grains, and plant oils and low ratios of intensely processed food sources high in sugar, salt, starch, and additives.

These new scientific shifts assist with explaining different vulnerabilities and controversies in modern nutrition. Following decades of the spotlight on basic, reductionist measurements like dietary fat, nutrient density, saturated fats, and energy density, the arising true complexities of various food sources and dietary models create genuine difficulties for figuring out effects on health and prosperity. For a few food varieties, significant quantities of planned observational or interventional reviews have become available just recently. Growing acknowledgement of the significance of dietary patterns has inspired scientific inquiry and a storm of experimental, business, and dietary examples of change formation and research backing (Corella & Ordovas, 2009). These variations, for instance, include flexitarian, vegetarian, low carb, gluten free and, paleo. A considerable number of these examples have specific goals (for example, weight loss, overall health, and anti-inflammation) and depend on contrasting understandings of current proof.

In poor nations, concerns on vitamin supplementation have arisen, for example, harms related to higher portion vitamin A dosage, risk of compounding infections like jungle fever with iron, and safety concerns on folic acid flour fortification, which could fuel neurological and mental deficits among individuals with low vitamin B12 levels. Furthermore, a sharp rise in non-genetic conditions in these nations has prompted a new spotlight on the "twofold burden" — both expectedly considered hunger (low calories and micronutrients) prompting poor maternal and child health and current lack of healthy sustenance causing type 2 diabetes, cardiovascular conditions, obesity, and cancer. These double universal burdens are progressively found within the same country, local community, family, and even individuals.

However, following decades of scientific discussion on global nutrition and supplements, food fortification, and starchy staples to offer calories, the vital shift towards diet quality is reduced by high inertia. This is considered, for instance, in the reductionist, single nutrient focal point of most UN sustainable development objectives (Deb et al., 2014). Even when non-transmittable disorders are considered, the overwhelming spotlight is on obesity as opposed to the different risk pathways and factors impacted by nutrition —facilitating a false idea of "over nutrition" as opposed to unhealthy dietary structure as the root cause.

4.9 NUTRITION PLAN

Similar to nutrition science, policies must move from brief reductionist methodologies to diverse methodologies. The nutrition strategy to decrease non-transferable diseases has up until this point commonly depended on consumer information — simply educate the public through schooling, dietary rules, food nutrition labels, and so on, and individuals can make better decisions. Notwithstanding, it is presently evident that data alone can't affect behavior, and that more extensive structures, policies, and ecological systems are required for successful change.

Intensifying these difficulties, different modern strategies are still focused on reductionist developments, for example, total fat or overall saturated fat, neglecting the significance of food variety and quality, processing strategies, and diet patterns. One more example of policy lag includes energy balance. Strategy developers continue promoting absolute calorie marking regulations for menus and packaging, including other calorie reduction approaches, as opposed to planning to increase calories from good food sources and lessen calories from bad food sources (Connolly & Dawson, 2015).

The public is naturally confused by constant advances in dietary messages. Different food organizations compound the alarm by marketing products containing refined sugar, flours, salt, and modern added additives, taking advantage of increased micronutrients or terms, for example, "organic," natural" or "local" to deliver a misleading sense of health. Public vulnerability is enhanced by changing nutritional messages from diverse media sources, digital and social networks, sociocultural leaders, and business outlets, whose messages differ based on underlying objectives, skills, points of view, and contending interests.

While reductionist strategies may have a value to lessen certain additives— e.g., sodium, added sugar and trans-fats — entire food-based strategies will be essential to completely address diet-based diseases. Most policy advances have zeroed in on sugar improved drinks, following the structure of the WHO Tobacco Control Framework Convention: duty, limited sale areas, limited marketing, use of warning tags, etc.

This factor disintegrates for incentivizing the use of quality food sources. Coordinated approaches, investments, and sociocultural techniques are expected to create changes in food production, workstations, schools, medical service structures, quality standards and design, food assistance projects, research and development, and public-private collaborations (Connolly & Dawson, 2015).

To be successful, future nutrition strategies should combine contemporary research advances on dietary needs (particular food sources, handling techniques, added substances, diet models) with trusted correspondence to the public and contemporary proof of viable structural level change. This consists of a shift from the universal determinants of health toward fulfilling the needs of an interconnected community, sociocultural, public, and worldwide determinants of food conditions and choices. In both low and high income nations, interventions should consider the double risks of food insecurity and terminal illness, and their connections to differences in education, revenue and opportunity (Wiwanitkit, 2012). This will need considerably more research funding, from government and private sources. Guided by information from previous studies, new methodologies are required to fast-track coordination of scientific studies and interpretation of modern and future developments.

4.10 SUGGESTIONS FOR MODERN RESEARCH, DIETARY RULES, AND FOOD STRATEGY

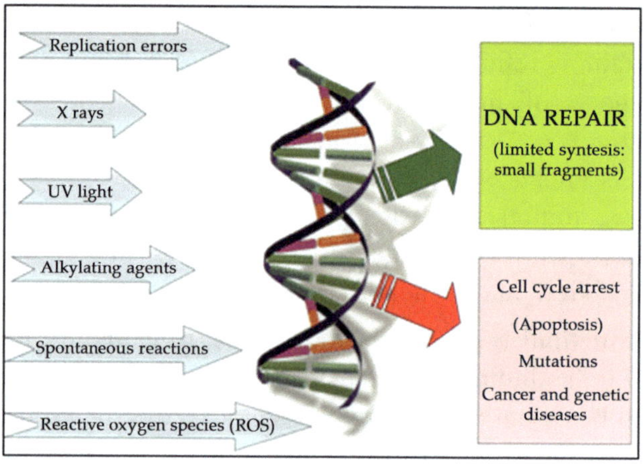

Figure 4.4: Nutrigenomics is used in cancer prevention.

Source: https://www.intechopen.com/chapters/44354

Through the review of gene expression, haplotypes, SNPs, and biochemical and physiological effects, researchers have confirmed personal and group differential reactions consume less carbs. This segment will assess chosen models of monogenic and composite, or polygenic, disorders, gene and eating routine association, and suggestions for the food sector. Genomic planning and cloning techniques have transformed the cycle of confining and recognizing genes of monogenic illness and prodded the utilization of comparative systems for evaluating genes associated with complex conditions and phenotypes (Lander, 1996). To date, around 1,000 human disorder genes have been identified and to some extent portrayed — 97% of these genes are now known to cause monogenic disorders (Jimenez-Sanchez, 2001). Common monogenic disorders are sickle cell anemia, caused by one amino hemoglobin change that reduces hemoglobin's oxygen propensity, and phenylketonuria, caused by a transformation in phenylalanine hydroxylase that prompts toxic amounts of phenylalanine.

In any case, the chemical basis of terminal illness isn't well recognized, despite the more than 600 related studies distributed to date (Hirschhorn et al., 2002), the failure to distinguish single genes susceptible to terminal diseases prompted the environmental variation disease debate (Lander, 1996; Collins et al., 1997), which is to a great extent liable for the genome-driven approach to the investigation of terminal diseases (Hirschhorn et al., 2002). This speculation, in its basic structure, blends naturally existing gene variations (i.e., alleles of unlinked genes) rather than changes causing any given terminal disorder.

An illustration of such gene variation is the Pro12Ala polymorphism in the peroxisome proliferator active receptor PPAR believed to cause type 2 diabetes. Expanding on evidence from complex impacts of various food sources, handling strategies, and diet patterns, new research needs are emerging in nutrition science. These incorporate ideal dietary strategies to decrease weight gain and obesity; associations between prebiotics and probiotics, matured food varieties, and stomach microbiota; impacts of unsaturated fats, flavonoids, and other bioactive molecules; custom nourishment, particularly for non-genetic factors, sociocultural, and microbiome factors; and the strong impacts of societal position on health and disease disparities.

For poor countries and populations, thorough research is needed to comprehend the ideal dietary models to mutually handle maternal health, child development, infection risk, and non-transmittable conditions. Our interpretation of diet-based biological pathways will keep on expanding, stressing the constraints of utilizing single surrogate results to decide the full health impacts of any dietary component. Likewise, future decisions about diets and health must be founded on corresponding proof from controlled interventions of diverse substitute endpoints, mechanical research, imminent observational investigations, and, when available, clinical tests of disease outcomes (Wittwer et al., 2011). This will need moving away from the modern oversimplified conviction that dependable dietary evidence can be gotten exclusively from large-scale randomized tests.

Considering the massive and continual rise in agribusiness and industrial food varieties, nutrition science should keep up with new developments in the food business and their health impacts. Fairly minimal rigorous assessment has been performed on possible long-term health impacts of present-day shifts in agrarian activities, animal breeding, crop reproduction, and food handling strategies, for example, grain processing and handling; plant oil extraction, deodorization, dairy fat homogenization; and application of emulsifiers and food thickeners.

Extra complexity may emerge in nutritional suggestions for general health versus treatment of unique circumstances. For instance, dietary recommendations for handling obesity are currently controversial. Different researchers continue supporting a fundamental "energy imbalance" idea of obesity, wherein calories from various food varieties are undeniably considered equal. Conversely, that's what developing proof recommends, over longer periods, diet synthesis might be a more significant emphasis than calories due to the changed impacts of various food sources on covering pathways for weight control, for example, fulfillment, mind reward, glycaemic reactions, the microbiome, and liver function. Across months and years, a few food varieties might impede pathways of weight homeostasis, while others might have generally impartial impacts, and promote the trustworthiness of weight guidelines (Van Breda et al., 2015). These long-term impacts will be particularly important in promoting optimal weight management (weight reduction in individuals with obesity) towards basic prevention (evasion of long-term weight gain).

Acknowledgment of complexity is a vital example of previous nutrigenomic studies. This creates space for scientific progress in nutrition, clinical medication, physical science, political science, or financial aspects: original observations led to improved knowledge and understanding of viable benefits, which are then definitely improved by new knowledge and acknowledgement of growing complexities.

4.11 SUMMARY

Figure 4.5: The nutrigenomics market is growing rapidly since its inception in 2001.

Source: https://www.medgadget.com/2020/01/nutrigenomics-market-experts-anticipate-market-boom-in-2026-major-players-basf-royal-dsm-danone.html

Nutrigenomics is a new science that examines a specific area of nutrition and applies sub-molecular instruments to investigate access and comprehend the different responses gotten through a specific diet routine applied among individual and population groups. The necessity for the utilization of customized nutrition in patients is growing and research is being made on its potential impacts.

Notwithstanding, research on nutrigenomics and specifically, obesity is as yet continuous. The phrase"nutrigenomics" was first defined in 2001 by Pelegrin (2001) and afterwards, it showed up in 2002 in a study by Van Ommen and Stierum (2002). A literature review on the PubMed database on using 'nutrigenomics' as a search phrase produced about 1005 articles starting around 2001 until the present. This search on its own shows fast progress into the future of nutrigenomics and the fields that should be investigated further. Customized nutrition will become the future when planning and prescribing an eating regimen for people based on their genome and their genetic variations. As per a review conducted in Greece on the general population (N = 1504) and health specialist (N = 87) samples, researchers have observed that there is an expanded requirement for the developing nutrigenomics research (Pavlidis et al., 2012). During the research period, just 11.5% of the subjects were encouraged to embrace a nutrigenomics test, suggesting that there is definitely not a massive application among nutritionists, clinical specialists or health specialists.

The paper is of 2014 and we are in 2022-hence, the word "recent" was removed

CHAPTER 5

BENEFITS OF NUTRIGENOMICS

CONTENTS

5.1 INTRODUCTION

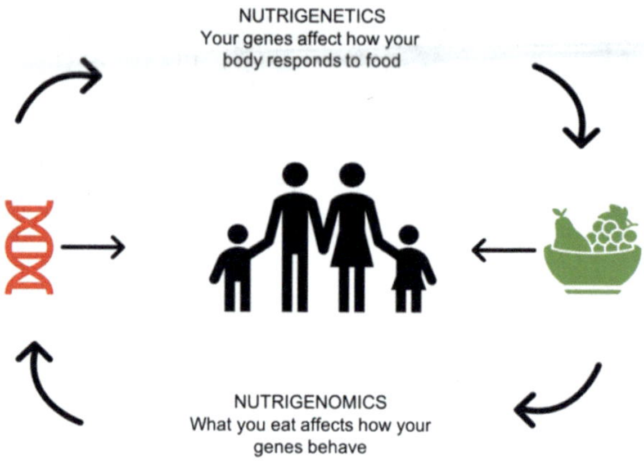

Figure 5.1: There are several benefits of nutrigenomics.

Source: Alvarez-Pitti J, de Blas A, Lurbe E. Innovations in Infant Feeding: Future Challenges and Opportunities in Obesity and Cardiometabolic Disease. Nutrients. 2020; 12(11):3508. https://doi.org/10.3390/nu12113508

Nutrigenomics makes use of the knowledge derived from scientific studies showing that nutrients in food can be used to generate changes in gene behavior. These studies also show that nutrients contribute to people's risk for cancer among other diseases. A scientific study of nutrients gives researchers hope of finding new ways of using food in preventing, curing and reducing the risk of diet-related disease and benefits include the increasing concern for one's health. It also gives nutritionists a chance to personalize nutrition and optimize food to ensure good health. It also enables them to discover genetic vulnerabilities that can be a storing motivating factor that encourages people to make the needed dietary and lifestyle changes (Su & Lu, 2012). There is a higher likelihood of people heeding to the advice given as they have already paid for it. Currently, for one to get their DNA profiled and analyzed it costs between $300 and $3,000. In most cases, large-scale food corporations spend large amounts of money on Nutrigenomics to enable them to develop enhanced or fortified products to deliver personalized diets.

The same applies to multi-national corporations that specialize in skin care, beauty products and aging. Nutrigenomics enables them to make very useful products. The benefits of Nutrigenomics can be derived from its aims. Nutrigenomics allow scientists to have personalized dietary regimen that encourages them to become more health conscious, more people are more likely to heed advice that they can pay for, it enables people to prevent or ameliorate diseases, and researchers are able to discover genetic susceptibilities that can become a strong motivator to making lifestyle and dietary changes. The other aim is to create safe upper and lower limits for essential macro-nutrients such as fats, carbohydrates, proteins and micronutrients including minerals and vitamins can be better understood and defined. It allows the extension of lifespan, it also allows people whose health is relatively unaffected by diet to continue enjoying the food they enjoy and finally it should enable people to avoid unnecessary vitamins and other dietary supplements. Nutrigenomics allows those with a commercial interest to respond to established market segments of early adopters seeking to develop new tools to be used to enhance health.

5.2 NUTRIGENOMIC ALLOWS GENES GUIDE OUR DIET

One of the general benefits of nutrigenomics is that it allows genes to dictate what people should eat. By focusing on the genes, bodies will be able to thrive. Cells have the ability to adapt to environmental changes through the regulation of gene expression that is essential for organism survival. There is a variation in gene expression between groups of people. This is in the presence or absence of nutrients through the increase or decrease in the production of cellular protein necessary for life sustaining functions (Singh & Sharma, 2019). This evolutionary process can be well explained by the development of gene mutation that alters the ability to tolerate lactose and adult mammals more so in those who are unable to digest lactose. The science of nutrigenomics is very useful as it promises to offer health practitioners get access to knowledge that enables them to predict potential genetic responses to nutritional intake. It also enables them to target and modify associated behavior.

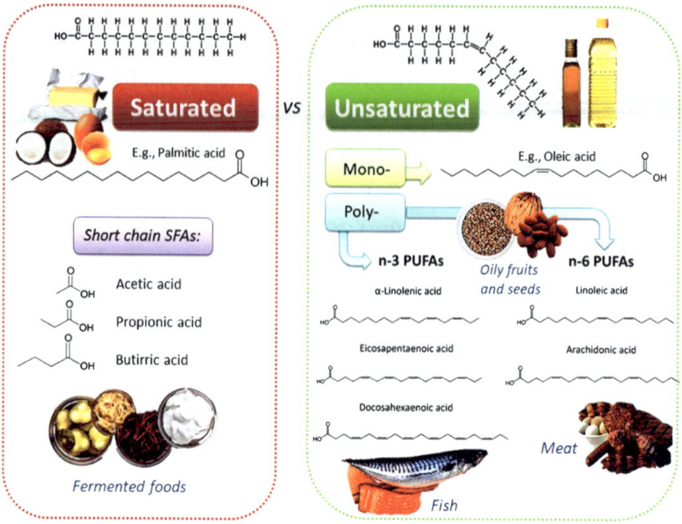

Figure 5.2: Nutrigenomics enables people to know the right food suited for their body.

Source: Bordoni L, Petracci I, Zhao F, Min W, Pierella E, Assmann TS, Martinez JA, Gabbianelli R. Nutrigenomics of Dietary Lipids. Antioxidants. 2021; 10(7):994. https://doi.org/10.3390/antiox10070994

This is very beneficial for various groups of people. Nutritional genomics is defined as the study if genes and nutrients interact in your genes while predicting the manner in which the body will respond to certain nutrients. Nutrigenomics places much focus on various genes more so the FTO gene. This gene is closely related to our body's energy expenditure, energy balance and metabolism. In the long run, these genes affect weight management and body composition. Variants in the FTO reveal how the body metabolizes protein and fats (Sikalidis, 2019). The dietitian is able to provide information that can be used by their clients in choosing an eating plan that works suitably with our genetic makeup. The results of nutrigenomic testing are used in making informed decisions. A good example is when genetic variants are able to signal an individual that they are more likely to develop high blood pressure or cholesterol. With this knowledge, one can take the needed steps to prevent the conditions from occurring. This can reduce the likelihood of sodium and saturated fats consumption.

If the body signals that an individual is likely to crave sweets, the individual can create a game plan that will counter their cravings so they avoid eating too much sugar. This means reduces the likelihood of an individual suffering from Type 2 diabetes. If the body signals that an individual is likely to get jitters when they drink caffeine, the individual knows the effects and is likely to feel inspired to skip coffee (Riscuta, 2016). When the body signals the likelihood of weight loss on high-protein, then the individual will confidently stick to an eating plan knowing how well their body will respond to macronutrients such as protein, fat and carbohydrates. If the genes dictate that the body will burn more fat with cardio or strength training then the individual should zero in on their most efficient system to burn fat at a faster rate.

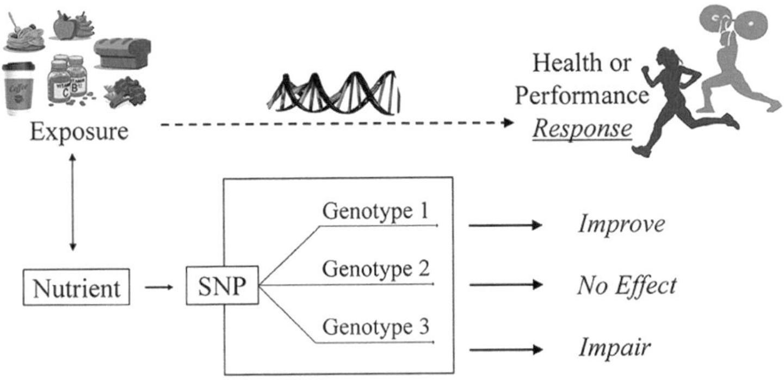

Figure 5.3: Nutrigenomics enables an individual to know workout routines that will be effective for their body.

Source: https://www.frontiersin.org/articles/10.3389/fnut.2019.00008/full

When using genes to dictate an individual's diet, nutritionists make use of nutrigenomic testing. This test works by conducting a DNA test for Nutrigenomics by conducting a simple swab of the inside of the mouth. The swab is then sent to a lab where it is analyzed and the results are generated after a few weeks. When in the lab, the swab is used to analyze 70 or more different genes. With regards to the accuracy of the test, if the test is done on a scientific basis then it is very accurate. These results keep in mind that the results are an indication of the potential for health conditions and the body's response to nutrients. It is important to note that even though a person is genetically probed to something does not mean it's a given. Other than this test, there is the over-the-counter nutrigenomic test. These tests are sold directly to consumers. However, when using such tests it is important for consumers to invoke the use of a professional. The relevance of the test is well explained by registered dietitians who have been well trained in nutrigenomics. The tests have different uses with regards to body health while ensuring that foods will minimize the risk on the basis of the specific variants. For instance, if an individual has a given genetic variant associated with high levels of triglycerides circulating in the blood (Riscuta & Dumitrescu, 2012). This can be reduced when the dietitian can look into various ways of adding more omega-3 fatty acids to the diet. This will be effective in lowering triglyceride levels.

5.3 NUTRIGENOMICS ALLOWS FOOD TO BE USED AS FOOD OR MEDICINE

There are various chemicals in food that can be utilized as medicine. Nutrigenomics entails the study of how naturally occurring chemicals in food can be used in altering the molecular expression of genetic information in every individual. In the real sense, the term nutrigenomics is used to refer to the effect of a person's diet on their gene expression or to the impact of genes on the response of a given dietary pattern, supplements or functional food on a specific health outcome. This was developed after the genomic era. This era can be described as the period in which there was a study on the relationship between diet lifestyle, gene and health that is nutrition regulate gene function like metabolism, translation and transcription such as diet gene interaction (Ray, 2014). Nutrigenomics is very useful in the use of food as medicine in that it places much focus on understanding nutritional influences on the maintenance and metabolism of internal equilibrium in the body and how vets it can be regulated depending on the diet and how it relates to the diseases. They also offer a powerful and exciting approach to unraveling the effects of diets on health.

Figure 5.4: Nutrigenomics allows easy identification of medicinal foods.

Source: https://blog.ginihealth.com/the-rise-of-nutrigenomics/

Initially, much emphasis was placed on nutritional research concentrated on nutritional deficiencies and impairment of health. However, venturing into nutrigenomics leads to the creation of a junction between genomics and health and allows the promotion of an increased understanding of how nutrition influenced metabolic pathways and control homeostasis.

During biomedical research in a private sector firm, the public and food industry recognized the need for nutrigenomics research that will be used as a basis for the development of the concept of personalized diet in the identification of molecular biomarkers. In the past few years, there has been a growing increase in the interest in nutrigenomics as a research topic. This is attributed to the fact that it is an area that has been viewed as worthy of public funding as a field with health care and commercialization possibilities and a basic scientific inquiry. The new understanding of scientific nutrigenomics had led to the increased commercial development of nutraceutical and functional goods that can be modified by the negative health effects of individual genetic profiles. Nutrigenomics has the main goal of improving dilatory advice, preventive strategies, reduction of healthcare costs and development of health promoting supplements (Qadeer & Ashhar, 2014). More research will need more patience, strong advocacy and realistic expectation for the needed research funding and the major focus of nutrition research on the prevention of chronic diseases such as cardiovascular disease, cancer and metabolic disorder.

Nutrigenomics is used in simply managing or treating diseases or the symptoms linked to diseases. It is also used in identifying susceptibilities to diseases allowing individuals to implement proactive measures to ensure that they avoid contracting the disease in the first place. For this reason, nutrigenomics in the future may lead to the development of evidence based healthful food and lifestyle advice. It could also lead to dietary intervention for contemporary humans. Currently, the advent of modern science has allowed researchers to conclude that not only are certain nutrition essential but also the specific quantity of each food type necessary for optimal health. This led to notions such as dilatory recommendations and nutritional epidemiology. It also led to the realization that food can directly contribute to the start of a disease. For this reason, human development to various diseases can be well defined by environmental influences such as diet, physical activity and smoking education. They can also be influenced by genetic elements that indicate that both aspects should be taken into consideration so as to optimize good health.

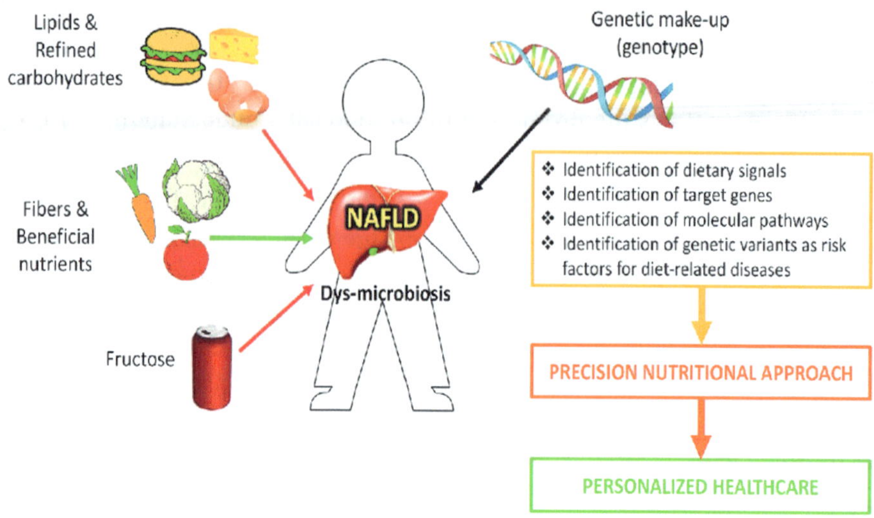

Figure 5.5: Knowledge of nutrigenomics has been used in the treatment of various diseases.

Source: Dongiovanni P, Valenti L. A Nutrigenomic Approach to Non-Alcoholic Fatty Liver Disease. International Journal of Molecular Sciences. 2017; 18(7):1534. https://doi.org/10.3390/ijms18071534

There is so much excitement with regards to nutrigenomics. The excitement is a result of the growing awareness of the potential for modification of diets or food in supporting human health while reducing the risk of diet-related diseases. This means that by identifying individual genetic predisposition for chronic diseases and the potential for individuals to respond to dietary intervention. This ensures that these diseases are well prevented by proper dietary intake. This means that nutrigenomics brings together the science of bioinformatics nutrition, molecular medicine and molecular biology genomics.

Usually, nutrigenomics involves the application of high-throughput genomics tools in the study of diet-gene interactions to identify dietetic components with several beneficial and detrimental health effects. Biomarkers were traditionally related to the onset of disease or organic damages were used in quantifying the effects. At this age, it is very essential to quantify phenotype changes that are close to or within the range of a healthy state. It also has a major focus on nutrient deficiencies and there is a relationship between health and nutrition (Pauletto et al., 2020). However, the development of genomics has allowed the creation of unprecedented opportunities that allow the generation of more information that can be used to create an understanding of nutrigenomic modulating genes and protein expression and ultimately has a big impact on cellular and organizational metabolism.

When studying nutrigenomics, it is important to note that it embodies three normative concepts. The first concept deals with food that is exclusively interpreted in terms of the prevention of diseases. The second concept deals with striving for health that can be used to interpret as the quantification of risks and prevention of diseases through positive food-gene interaction. The final concept makes use of the normative idea that the prevention of diseases can help minimize the risk as the responsibility of the individual. Currently, it is said that nutritional factors cause about 30-60% of cardiovascular diseases, diabetes, cancer and other diseases most of which are rapidly increasing (Peña-Romero et al., 2018). This has led to the development of a new branch of genomic research that is best summarized by five tenets of nutrigenomics. The five tenets are as follows:

In normal circumstances and among certain groups of individuals, the diet they take can pose a serious risk factor for a number of diseases. In the second tenet, common dietary chemicals can act on human genomes either directly or indirectly so as to modify gene expression or structure. They also influence the degree to which diet influences the balance between healthy and disease states depending on the genetic make up of the individual. Some diets have highly regulated genes such that they are able to affect the start, incidence, progression and severity of chronic diseases. The use of knowledge of nutritional requirement, genotype and nutritional status through dietary intervention can be used in the prevention, migration and curing of chronic diseases. For this reason, genomics is considered significant in the matter of improving public health. Its effective utilization has made this kind of research more widespread. It affects the manner in which nutritionists and the general public understand and practice nutrition.

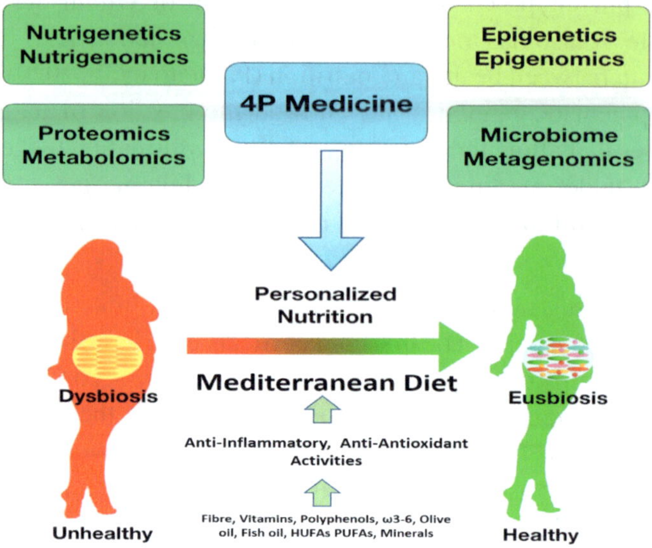

Figure 5.6: Nutrigenomics allows the identification of foods suitable for people in a given age group.

Source: Di Renzo L, Gualtieri P, Romano L, Marrone G, Noce A, Pujia A, Perrone MA, Aiello V, Colica C, De Lorenzo A. Role of Personalized Nutrition in Chronic-Degenerative Diseases. Nutrients. 2019; 11(8):1707. https://doi.org/10.3390/nu11081707

When using food as medicine, there are questions between genetics and life choices. In this regard, there is a question of genetics as one part of the health picture. It also takes into consideration other factors including the person's environment and lifestyle and their effects. There are different ways in which all the elements contribute to the manner in which the body responds to food. There is a study involving about 1,000 participants who generated data. About half of the participants were twins. (Piroddi et al., 2017). Afterward, their blood sugar level was measured. Other elements measured include fat and insulin in the blood. In the study, information from each of the participant's amount of exercise and sleep was collected. Stool samples were also obtained. Data was also obtained from the participant's gut bacteria.

The findings include the fact that meal macronutrients such as carbs, fat and proteins had the biggest impact on sugar levels in the blood. It was also established that blood-lipid levels are mostly affected by gut bacteria. It was also noted that genetics had a great effect on blood-sugar levels compared to lipids. Its effects are less significant when compared to meal composition. This generally means that genetics play a vital role in complex dance among factors varying from one person to the other. Some of these factors including gut bacteria, diet, exercise and sleep can be modified.

5.4 NUTRIGENOMICS EXPLAINS OMEGA-3'S IMMUNE HEALTH BENEFITS

There are various components found in food. Among them is omega-3 found in certain foods such as fish. Omega-3 fatty acids are known to lower LDL cholesterol and raise good HDL cholesterol and protect the body from certain types of cancers. It also helps protect the body from diseases, high blood pressure, degenerative eye disease and arthritis. Omega-3 can be found in various foods including flax seeds, canola oil and walnuts (Pal, 2018). The best kind of omega-3 can be found in marine fish. Within omega-3, there is a certain kind of fatty acid known as eicosapentaenoic acid. This fatty acid has been said to reduce the level of inflammatory genes in arthritic canine cells.

Figure 5.7: Nutrigenomics highlights the various uses and benefits of omega-3.

Source: https://www.verywellhealth.com/omega-3-fatty-acids-5216288

A study on the body shows that high levels of omega-3 are in the brain and are very useful in brain memory, behavior function and performance. It is worth noting that infants who do not receive enough omega-3 fatty acids from their mothers during pregnancy are at a higher risk of developing nerve and vision problems. Some symptoms of omega-3 deficiency include poor circulation, depression, mood swings, heart problems dry skin, poor memory and fatigue. Nutrigenomics has been key in highlighting the importance of omega-3 among various groups of people. According to these studies, there is a need of a proper ratio of omega-3 and omega-6 in the diet. This is attributed to the fact that omega-3 is useful in reducing inflammation. However, omega-6 fatty acids have a tendency to promote inflammation (Neeha & Kinth, 2013). Some of the advantages of Omega-3 as demonstrated by nutrigenomics include fighting depression and anxiety.

Currently, depression is one of the most common mental disorders affecting large groups of people across the world. Some of the symptoms include a general interest in life, lethargy and sadness. Anxiety also falls in the same group as depression. It is characterized by nervousness and constant worry. Though most people opt to go for therapy to deal with this issue, data in nutrigenomics shows that people who consume omega-3 are less likely to suffer from depression. It was also noted that when people suffering from anxiety or depression start taking omega-3 supplements or foods rich in omega-3, their symptoms improve. There are three types of omega-3 fatty acids namely DHA, EPA and ALA. According to nutrigenomics, EPA is the best kind of fatty acid that can be used in fighting depression. Studies have shown that EPA had been effective in fighting against depression. For this reason, it is a common anti-depressant drug.

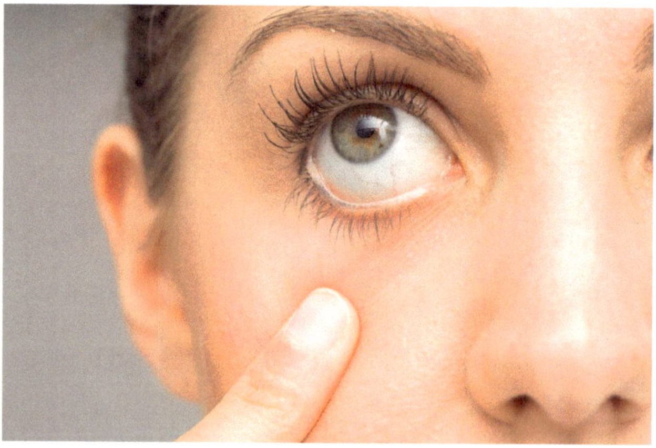

Figure 5.8: Omega-3 helps to treat dry eye.

Source: https://stoneycreekeyecare.com/how-omega-3s-can-help-treat-dry-eye/

Nutrigenomics has also proven that omega-3 can be used to improve eye health. Among the three groups of omega-3, the DHA type of omega-3 is considered a major structural component of the retina of a person's eye. It was noted that among people, vision problems were common for those who did not get enough DHA. However, it was noted that among the groups of people who got sufficient amounts of omega-3 had a reduced risk of macular degeneration which is one of the world's leading causes of permanent blindness and eye damage (McNamara, 2010). Other than that it allows the promotion of brain health during pregnancy and the early life of the child.

Figure 5.9: Infants given omega-3 have better health compared to their counterparts.

Source: https://omega3innovations.com/blog/fish-oil-for-kids-3-things-to-consider/

Nutrigenomics has also been useful in providing data showing that omega-3 is useful for the development and growth of infants. Its data shows that DHA accounts for 40% of the polyunsaturated fatty acid in a person's brain and the other 60% in the retina of the eye. For this reason, it is no surprise when infants are fed a DHA-fortified formula will have impeccable eyesight when compared to those that have not been fed the formula. In most cases, nutritionists and dieticians will include large amounts of omega-3 in a pregnant woman's diet. This will allow the child to benefit from the components. Such children will have decreased risks of ADHD, cerebral palsy, autism, decreased risk of developmental delay, fewer behavioral problems, better social and communication skills and higher intelligence.

Data from nutrigenomics has been useful in proving that omega-3 can be useful in improving risk factors for heart disease. Among the diseases causing death across the world are strokes and heart attacks. Studies among fish-eating communities recorded low rates of such diseases. Nutrigenomics links this to the consumption of omega-3. This has led to omega-3 being linked to numerous benefits for good heart health. Some of these benefits include the fact that omega-3 can get useful in reducing triglycerides usually in the range of 15-30%, with regards to blood pressure omega-3 can be used in reducing blood pressure levels in people suffering from high blood pressure (Mustapa et al., 2020). Omega-3 has been useful in raising good HDL cholesterol in the body, it is useful in managing blood clots by keeping blood platelets from clumping together. This is very useful in ensuring harmful blood clots are not formed in the body. This allows the body to quickly heal from any wounds.

Nutrigenomics can also prove that omega-3 can help keep protect the body from plaques so as to ensure that the arteries are free from damage and are smooth. The occurrence of plaque is very dangerous as it can harden or restrict arteries. Data can prove that omega-3 can protect the body from inflammation through the production of substances that counteracts inflammation. Based on the genetic composition of a person's body, omega-3 can be useful in lowering bad LDL cholesterol. Nutrigenomics also proves that omega-3 can help reduce symptoms of ADHD among children. ADHA is anattention deficit hyperactivity disorder that is a behavioral disorder characterized by impulsivity, hyperactivity and inattention. Large numbers of children suffering from ADHD tend to have low levels of omega-3 in their blood compared to healthy peers. Research done in nutrigenomics can prove that omega-3 supplements can help lower symptoms of ADHD. This means that such children will have improved ability to be attentive in class and task completion. It is useful in reducing aggression, restlessness, impulsiveness and hyperactivity. Recent data shows that the consumption of fish oil supplements produces promising results when treating ADHD.

Nutrigenomics provides evidence to reduce symptoms of metabolic syndrome. Metabolic syndrome is defined as a collection of conditions. It mostly takes into consideration central obesity commonly known as belly fat, high blood pressure, low good HDL cholesterol, high triglycerides and insulin resistance. This kind of disorder is a major public health concern as it leads to increased risks of many other illnesses such as diabetes and heart disease. Consumption of omega-3 in the diet has been useful in improving heart disease risk factors, inflammation and insulin resistance among people suffering from metabolic syndrome (Masotti et al., 2010).

When fighting inflammation, omega-3 helps deal with infections leading to infections. Inflammation is usually the body's response to infections and damage in the body. This means that inflammation is vital for health. There are cases where inflammation persists for longer durations of time even in the absence of injury and infection. This kind of inflammation is referred to as chronic or long-term inflammation. This kind of inflammation usually contributes to almost every chronic western illness as the likes of cancer and heart disease. Information from nutrigenomics shows that omega-3 fatty acids are useful in reducing the production of molecules and substances linked to inflammation such as inflammatory cytokines and eicosanoids. There is sufficient information to prove that there is a close link between reduced inflammation and high omega-3 intake.

Nutrigenomics also shows that omega-3 is useful in fighting autoimmune diseases. Autoimmune diseases are characterized by the body becoming highly reactive after mistaking healthy cells for foreign cells and therefore start attacking them. A good example is type 1 diabetes where the body's immune system attacks insulting-producing cells in the pancreas. Omega-3 is very useful in combating most of these diseases more so during early life (Mathers, 2017). Enough supply of Omega-3 during the first year of life is linked to low occurrences of autoimmune diseases including multiple sclerosis, autoimmune diabetes and type 1 diabetes. Other diseases that can be treated by omega-3 include lupus, psoriasis, Crohn's disease, ulcerative colitis abs rheumatoid arthritis.

5.5 NUTRIGENOMICS SHOWS BENEFITS OF COCOA IN BLOOD PRESSURE

Studies in nutrigenomics have been useful in proving that cocoa is useful in reducing high blood pressure. Cocoa has polyphenol compounds said to reduce blood pressure in certain genotypes. Polyphenols have an activity known as antiotensin-converting enzyme. This enzyme is usually a target in blood pressure medication mostly found in dark chocolate made up of 72% cocoa. This is usually accompanied by the degree of inhibition dependent on the genotype of the individual. All the ACE inhibitors work by inhibiting the conversion of angiotensin-I to the potent vasoconstrictor. This means that angiotensin-II is useful in improving blood pressure and blood flow (Loor et al., 2015). To know how cocoa works in the body, and its chemical composition will provide a lot of information. Cocoa is made up of various chemicals including antioxidants known as flavonoids. Though their role in the body is not well defined, they appear to cause relaxation in human veins. This reduces blood pressure.

Figure 5.10: Cocoa helps control blood pressure.

Source: https://www.shutterstock.com/search/cocoa

Cocoa is very effective in managing high blood pressure as research shows that those who eat dark chocolate and products of cocoa for 2-17 weeks can lower the top number in blood pressure reading also known as systolic blood pressure by 2.8-4.7 mmHg. It is also useful in lowering the diastolic blood pressure by 1.9-2.8 mmHg in people known to have normal blood pressure and high blood pressure. However, consumption of Cocoa is ineffective for high cholesterol. It is suggested that the consumption of cocoa products may not be effective in improving cholesterol levels among people suffering from high cholesterol. More research is needed in proving that cocoa is useful in treating heart disease. Little information could prove that healthy elderly men that consume large amounts of cocoa from dietary sources have reduced average blood pressure compared to those to those who take reduced amounts (Ladeira et al., 2018). The person's gene also influences the impacts of cocoa in the body wheresome people will have lower risks of death from heart disease. Consumption of chocolate and cocoa has been useful in improving the function of the inner lining (endothelium) of blood vessels which in the long run reduces the risk of heart disease.

5.6 NUTRIGENOMICS SHOWS BENEFITS OF MAGNESIUM METABOLIC ACTIONS

There are various nutrients used by the body that are derived from food. Magnesium is among those nutrients. Magnesium is involved in the up and down-regulate a number of genes linked to metabolism. They also have favorable effects on some metabolic pathways associated with changes in gene expression. Studies show that magnesium supplementation has been greatly associated with reduced levels of C-peptide. This is a marker linked to improved sensitivity to insulin. Magnesium has also been linked to the down-regulation of certain genes related to metabolic and inflammatory pathways. There are reports written in terms of gene expression that indicate that with regards to magnesium metabolic reactions, about 24 genes were up-regulated while 36 genes were down-regulated. These kinds of regulations are usually a response to magnesium supplementation (Kussmann & Affolter, 2009). Some records show that there is a systemic effect of magnesium supplementation that brings about measurable physiologic changes in the urinary proteome after treatment with magnesium for a duration of about 4 weeks. This means that there is a need for more investigation into such changes and identification of the proteins involved.

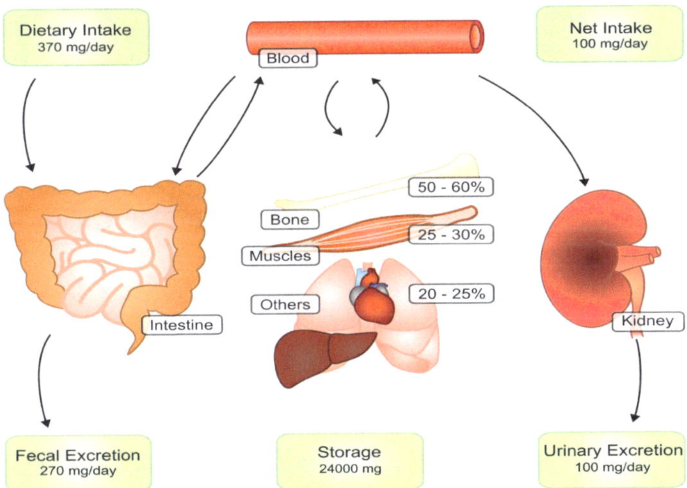

Figure 5.11: Magnesium is involved in metabolic actions.

Source: https://journals.physiology.org/doi/full/10.1152/physrev.00012.2014

5.7 NUTRIGENOMICS SUPPORTS EVIDENCE FOR HEALTH BENEFITS OF ANTHOCYANINS

Another benefit of nutrigenomics is that it provides vital information with regard to anthocyanins. Anthocyanins are described as a large sub-group of flavonoids mostly found in vegetables and fruits. These components are said to be safe and potent antioxidants. Such antioxidants exhibit diverse potential health benefits. These properties include anti-inflammatory properties, anti-diabetic, anti-cancer, anti-atherosclerotic and cardio protection. These antioxidants have the ability to cross the blood-brain barrier and finally distribute in the central nervous system. There are studies whose results have been useful in providing information on anthocyanins. This information indicates that they represent novel neuroprotective agents and are beneficial in ameliorating ethanol neurotoxicity (Kussmann & Van Bladeren, 2011). Currently, there is information that can prove that anthocyanins are widespread pigments in the plant kingdom. They have the potency of anti-obesity in mice. They also have the ability to enhance adipocytokine secretion and expression of the adipocyte gene in adipocytes.

Figure 5.12: Anthocyanins are found in fruits and vegetables.

Source: Khan, Muhammad & Ikram, Muhammad & Park, Tae & Kim, Myeong. (2021). Pathology, Risk Factors, and Oxidative Damage Related to Type 2 Diabetes-Mediated Alzheimer's Disease and the Rescuing Effects of the Potent Antioxidant Anthocyanin. Oxidative medicine and cellular longevity. 2021. 4051207. 10.1155/2021/4051207.

5.8 NUTRIGENOMICS COULD PROVIDE NUTRITION-RELEVANT BIOMARKERS

Nutrigenomics as a field along with nutrigenetics is very useful as they present a very powerful source of information that can be used when creating and implementing personalized diets. They are both useful as they specify the manner in which each individual is affected by genetic variance impacted response to the kind of nutrients they consume. This means that the individual will be able to avoid some implications of negative metabolic consequences through the selection and eating of the most appropriate foods. This is well demonstrated by people suffering from lactose intolerance wherethe body of the individual is unable to tolerate persistent allele and is not able to digest it. For such people, their health would benefit if the individual would refrain from consuming such kinds of foods. On the other hand, nutrigenomics looks into the manner in which the nutritional state of the body affects gene expression (Kang, 2012). This means that consuming certain nutrients according to the individual's health condition may normalize their disease-related gene expression. A good illustration is when a person exhibits increased systemic inflammation linked to an up-regulation of inflammatory cytokines that may be suppressed in the presence of certain nutrients.

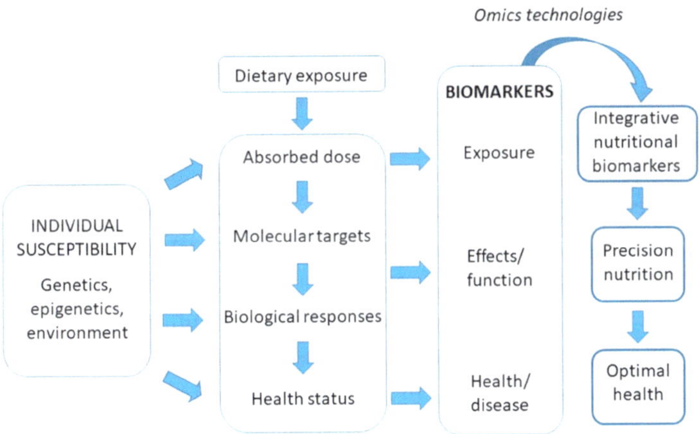

Figure 5.13: Biomarkers are used in nutrigenomics to generate personal diets.

Source: Picó C, Serra F, Rodríguez AM, Keijer J, Palou A. Biomarkers of Nutrition and Health: New Tools for New Approaches. Nutrients. 2019; 11(5):1092. https://doi.org/10.3390/nu11051092

When linking the two fields, the main concept is the point that such nutrients will be able to interact with an individual's genetic profile to improve their health. This means that each individual has their own genetic profile all of which is used when making personalized diets. This is very useful with regards to personalized nutrition which is the ability to precisely detect and interpret a person's individual metabolic and health state through these metabolic biomarkers. Such biomarkers are usually linked to molecular signatures linked to metabolism pathways. This takes into consideration genes or proteins responsible for the metabolism of nutrients as well as the metabolites derived from them. This is well illustrated when biomarkers of diabetes that include dysregulated insulin expression and elevated levels of blood glucose are noted in disordered glucose metabolism (Irimie et al., 2019). This means that biomarkers are very important in the development of personalized nutrition more so in areas that are linked to disease or pathological conditions, areas that are easily measurable such as urine samples or plasma and finally modifiable nutrient intake. Some of these biomarkers are a bit rare and may need great efforts being directed towards them.

It is advised that an integrated approach should be highly emphasized in the identification of metabolic biomarkers. Recently, the identification of biomarkers has focused on singular molecular species more so mRNA, metabolites or proteins. All these measurements may be used in showing one view of a particular pathway. When these methods are fully integrated they can provide a dynamic and full perspective for the interaction between nutrients and genes. Any alterations in metabolic pathways tend to involve changes in gene expression level and in metabolic content. Any changes in genes can be detected through the analysis of the gene expression such as proteomics or next-generation sequencing. Metabolic analysis is used in detecting metabolite changes (Hurlimann et al., 2014). A good example is the next-generation sequencing that identifies gene alternations in a particular pathway that includes increased COX-2 expression in eicosanoid formation. Metabolomic analysis can be used to indicate levels of corresponding metabolites in the pathway. This may include increased content of prostaglandins E2.

Combining this information together allows researchers and nutritionists to confirm that eicosanoid production is augmented. This is well demonstrated by the combination of such technologies to allow the detection of changes in both genes and nutrients. Such information gives nutritionists a comprehensive picture of metabolic pathways and the network between key elements. This works while reducing confounding factors while increasing accuracy during the identification of metabolic biomarkers. When establishing personalized nutrition, a great deal of work is still pending. New technologies can be harnessed in integrative ways that can allow nutritionists quickly and accurately identify an individual's nutritional needs. Such kind of knowledge can be utilized as a power adjunct to current health interventions. Currently, it is a major priority for nutritionists and researchers while developing new metabolic biomarkers that can easily be detected and respond to nutritional interventions. This goal can be achieved through research in nutrigenomics and nutrigenetics.

Nutrigenomics has also been useful in providing information on nutrition-relevant biomarkers. This means that any changes to messenger RNA and the corresponding proteins have been useful in control of the transport of certain metabolites and nutrients in a biochemical pathway (Alagawany et al., 2022). Studies in nutrigenomics have been useful in providing a new set of biomarkers with regard to nutrition.

5.9 NUTRIGENOMICS IS BENEFICIAL FOR OUR SKIN

Skin manufacturing companies have made use of nutrigenomics when manufacturing various products. There are various skin-related problems caused by the diet people use and inadequacies in their nutrition. Some of these problems include premature aging of the skin, dry skin, psoriasis, eczema and acne. The inadequacies in nutrition and poor diet have substantially contributed to the deterioration of such kinds of conditions and vice versa. Skin can be well managed by proper diet (Hesketh, 2013). A proper diet can improve the appearance of the skin and its health. In most cases, diets are chosen depending on the blood type and genotype of the individual. They may be made up of minimally processed seeds, nuts, legumes, vegetables and fruits.

Such diets could also be made up of fermented products obtained from unpasteurized and non-homogenized milk. Most of these foods are packed with nutrients needed for healthy skin such as vitamin E and B. Other than vitamins, it could contain minerals such as manganese, copper, iron, potassium, magnesium and calcium. In people with suitable blood groups, they are able to incorporate walnuts, almonds and flax seeds. Some fruits that can be eaten by people of all blood types include cranberries, raspberries, blueberries and pineapples. Meat is also a good addition to the diet. A Nutrigenomic diet profile can be used by an individual to get healthy skin. There are various products made using nutrients to help maintain good skin. Such products may include vitamin C serum among other products.

Figure 5.14: Nutrigenomics guides the therapist in healing skin.

Source: https://vitagene.com/blog/nutrigenomics-and-skin-aging/

When looking at the use of nutrigenomics in making skin care products, observations of the client's genetic makeup could prove useful in indicating any problems related to skin care. This kind of information is mostly used by skin therapists and other groups of people working in the skin care industry. There are a number of researchers who developed an interest in nutrigenomics in the field of skincare and work with skin care companies. Some develop an interest after an experience in genetic testing (Afman & Müller, 2006). This kind of experience allows them to get the information they so wanted and allowed them to venture further into such kind of research. There are various institutions known to offer a master's in the field of nutrigenomics. Such studies allow them to learn the interaction between skin health and genetics.

Currently, the use of nutrigenomics in skincare is still new. For this reason, it may be convenient and trailblazing. It is very useful in that it allows clinics and beauty salons are able to offer a truly personalized experience. Nutrigenomics has been key to demonstrating the relationship between good skin and consumption of nutrients. This kind of approach is considered the two-pronged approach. This approach looks into the fact that our genes can interrupt or interfere with certain mechanisms. This means that the skin is not always benefiting from all the foods a person eats. From this information, one can reduce the potential for aging and wrinkles by providing products or stimulating the collagen products that contain it.

This means that the person's genetics will tell them whether the food will be efficient as the person expects it to be. This gives an individual the ability to properly decide on products that could help improve the skin of the client from a skincare perspective. However, it is important to note that there environmental factors affecting skin as there are gene factors affecting the skin. For instance, if one sleeps with makeup on their face and wakes up with congested skin then these effects are not a result of genetic factors but environmental factors. Nutrigenomic looks at skin condition from the inside. This is because, it places much emphasis on the identification of weakness within the genes and combining treatments (Banerjee et al., 2015). Such kinds of treatments are referred to as 360- degree treatments. It takes into consideration the feel good on the inside to look good on the outside approach. This means that an individual has to nurture their DNA since the DNA cannot be changed. The only thing one can do is to support their DNA to ensure that they are in the best condition possible as it will be demonstrated in the well-being of the person and the condition of their skin.

Nutrigenomics is a useful tool for skin therapists as it contributes to the tools they can use to better improve their work. It enables them to know the kind of treatments they should use on their clients depending on their needs. There are cases where therapists create a treatment program for a client and on the retail side, they can improve their products that have been certified through genetics will improve skin conditions. It is important for the therapist to know that they need to use nutrigenomics to their advantage. This can be achieved through team gene skin profiling where one conducts an analysis of a swab obtained from the client. The analysis will reveal one of three genetic types. The first genetic type is of strength and is an indication of collagen production. The second deals with nourished genetic type that gives an indication that the body is in need of minerals and vitamins. This gene type is also used by other parts of the body before it helps improve skin. The final gene type of combination indicates that an individual needs both vitamins and collagen (Gonzalez et al., 2015). If the therapist is able to identify what the client lacks in their body they should incorporate it in their routine through supplements or skincare products.

There are companies that send testing kits to salons. This allows them to collect swabs from clients and send the swabs to the company for processing. After processing, the results are sent back to the salon where they can make the products or treatment programs in a personalized manner to cater to the needs of the client.

5.10 NUTRIGENOMICS IS USED IN HEALTH ECONOMICS IN WEIGHT MANAGEMENT

Many people are becoming more conscious about what they eat. This highlights the benefits of nutrigenomics in health economics more so with regard to weight management. In nutrigenomics, there is a theoretical modeling study where the main focus is to evaluate the health and economic implications of a Nutrigenomic product to deal with weight loss. The constructed nutrigenomic economic model was made by using the link present between published study data related to product efficiency and ingredients and validated clinical assessments that were initially tied to data on health economics that takes into consideration the condition prevalence and the general cost of illness (Guéant et al., 2014). When using the theoretical model, the demonstration is with the LG839 variant known to positively reduce the cost of illness at the microeconomic and macroeconomic levels on the basis of cost-effectiveness and cost-benefit analysis. This has allowed the prediction of health economic implications of a nutrigenomic intervention to vest demonstrate the theoretical model of nutrigenomic economics. Such a study is said to be hypothesis-generating and can be used in the definition of protocols when prospectively testing the health and economic benefits of nutrigenomics.

Figure 5.15: Nutrigenomics is useful in weight loss.

Source: https://www.instituteofnutrigenetics.in/blog-details.php?slug_url=know-your-genes-and-choose-your-sport/

With weight management, nutrigenomics is used to manage obesity. Obesity is a condition caused by multiple factors such as genetics, behavior and environment. The CDC conducted studies in 2006 that led to the discovery that about 50 genes are associated with obesity. Among the most common obesity-associated with obesity include MC4R and FTO. Fat mass and obesity-associated are coded by the FTO gene. This means that there is a strong relationship between obesity and risk alleles. There are a number of studies that can support the notion that the gene's risk got obesity bears a relation to effects on energy and satiety intake. On the other hand, the MC4R gene codes for the melanicortin-4-receptor found in the brain (Fenech et al., 2011). This gene makes use of a melanocortin system which is one of the ways in which the body is able to regulate body weight and feeding behavior, through satiety and hunger signals to the brain. In most cases, the MC4R risk alleles are linked to obesity, more food cravings and reduced satiety.

With regards to weight loss, nutrigenomics takes the understanding of obesity and genetics a step further by taking into consideration the manner in which diet interacts with obesity genes. Nutrigenomics can be used in directing dieticians on the needed modifications that can improve weight loss in those who have genetic risk alleles for obesity. A number of studies shows how the individual with a specific risk allele of the FTO gene is able to benefit from a high protein diet (Ferguson et al., 2010). There was a study that lasted 2 years that showed that carriers of the risk allele had a greater reduction in fat and weight distribution in response to a high-protein diet compared to individuals that do not have a risk allele. This means that people found with the MC4R risk can benefit from nutrition counseling that focuses on mitigation of food cravings and hunger. Often at times, it includes the practice of mindful eating and a diet rich in protein and healthy fats when promoting feelings of fullness and satiety.

Nutrigenomics also has an effect on human behavior that contributes to weight management. Most dietitians have knowledge of these effects and make use of it to help people with weight problems. When they make use of nutrigenomic testing in their practice, their clients become more receptive to recommendations and are motivated to stick to their diets. They are able to back up their claims with research. Research indicates that the use of nutrigenomic testing in diet personalization is useful in leading to better compliance and longer-term weight loss and diet. Dietitians should note that there are proven benefits of the use of nutrigenomics in weight loss (Benítez et al., 2017). The Academy of Nutrition and Dietetics recently released a report that accurately states that genetic testing should be used as a part of nutrition assessment when giving a holistic, personalized diet intervention. It is advised that these tests should be made up of well-researched genes. As a new science, it is important to note that not all genes have been used in nutrigenomics as not all of them are useful.

5.11 NUTRIGENIC ASSOCIATION OF THE 5-LIPOXYGENASE GENE WITH MYOCARDIAL INFARCTION

The rate-limiting step of the biosynthesis of proinflammatory leukotrienes is catalyzed by 5-Lipoxygenase from arachidonic acid. The 5-Lipoxygenase is from arachidonic acid and has been associated with atherosclerosis in human and animal models. There are reports that indicate that variants of 5-LO promoter repeat polymorphism have been linked with carotid atherosclerosis in humans (Ferguson, 2016). The generated effect was exacerbated by high dietary amino acids through high dietary N-3 fatty acids.

CHAPTER 6

DRAWBACKS OF NUTRIGENOMICS

CONTENTS

Several ethical considerations must be addressed, including the role of stress in disease development, individual privacy threats, and insurance companies exploiting nutrigenetic data to refuse coverage to patients with particular conditions. Because nutrigenomics-based test kits are expensive, they may not be feasible in poor countries such as India, where the majority of the population cannot afford them. It must be accessible and affordable for a large portion of the population, according to current public health and disease management practices. Furthermore, determining what nutrigenomics information and services are required in developing countries for efficient and cost-effective adoption is difficult. It's also uncertain if it will be able to keep its promises and allay everyone's fears, and perhaps most crucially, whether the food it creates will be food (Fenech, 2008). Traditional diets and the true significance of food will be hard to sustain under these conditions. Nutrigenomics-based advice should not be broad but rather tailored to the individual demands of certain regions and populations. It is also critical that the testing fees are reasonable and comparable to the average household income. To perform a comprehensive study, researchers should concentrate on the typical diets encountered in both industrialized and developing nations, with only minor alterations to the RDA.

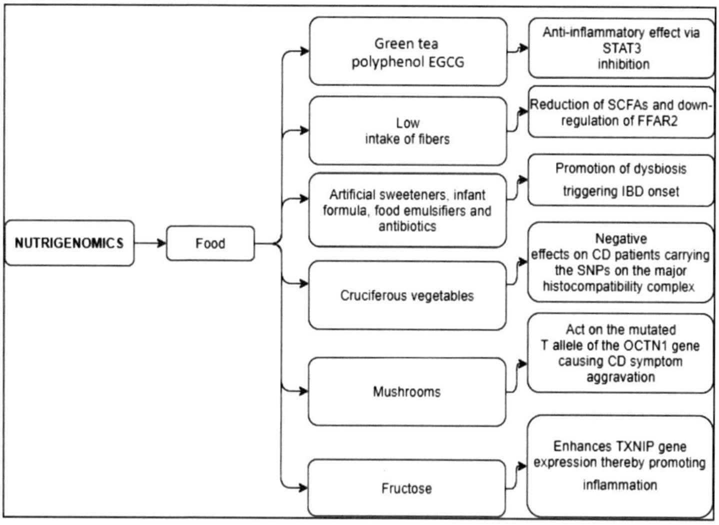

Figure 6.1: Nutrigenomics in Crohn's disease.

Source: Caio G, Lungaro L, Caputo F, Zoli E, Giancola F, Chiarioni G, De Giorgio R, Zoli G. Nutritional Treatment in Crohn's Disease. Nutrients. 2021; 13(5):1628. https://doi.org/10.3390/nu13051628

Despite the expansion of nutrigenomics research, applications and implementation are still in their infancy. More money, effort, and detailed molecular data are required to better understand the onset of illness, individual genetic makeup, and dietary differences. The bulk of the "pros and drawbacks" of nutrigenomics, like any other scientific discipline, have to do with premature health claims or validation, the financial interests of sponsors, and ethical quandaries (Fallaize et al., 2013). The credibility of a promise may be jeopardized if the most fundamental difficulties and concerns are not addressed. Variances in the amount and type of nutrients and bioactive compounds in diverse meals, as well as differences in food composition and preparation techniques, are obstacles to adopting molecular nutrition theories and establishing personalized diets for a given population.

Furthermore, it is unknown how existing drugs and dietary supplements interact with one another. When utilizing methodologies and detecting "subtle" biological reactions to bioactive foods, care must be taken. To comprehend the interdependent effects of bioactive chemicals on global gene expression, cell function, and individual health status, a multidisciplinary approach based on system biology is required. For the most successful implementation of nutrigenomics, bioinformatics tools for data processing, clustering, and integration at several 'omics' levels must be developed. Some chronic illnesses and health conditions may only be recognized early if a nutritional signature is created, which is why a single biomarker is insufficient. To gain public acceptability and utility, cutting-edge analytical tools, data analysis software, and cost-effective processing must be used. It's also challenging to review, interpret, and match data from different platforms, much alone assure the repeatability and validity of known biomarkers.

Nutrigenomics has certain disadvantages, including the fact that it is expensive and time-consuming.

To discover the causal relationship between nutrition and illness prevention and outcome, nutrigenomic research demands costly and time-consuming long-term intervention trials. Nutritional assessments, for example, are less exact than genetic or biochemical tests. Another disadvantage of nutrigenomics is its racial implications. Because nutrigenomics incorporates genetic testing on samples from defined populations, there may be ethical and legal concerns. When corporations and insurance firms exploit genetic information for non-medical purposes, ethical concerns arise (Elsamanoudy et al., 2016).

Figure 6.2: A healthy breakfast should have multiple nutrients.

Source: https://www.issaonline.com/blog/post/nutrigenetics-vs-nutrigenomics-what-they-mean-to-clients

Some businesses may also take advantage of the nutrigenomics sector.

American businesses are capitalizing on the increased public interest in nutrigenomics by making their services available to the general population. These companies may provide medically dubious health-related forecasts and expensive DNA supplements to mislead their clientele. Nutrigenomics tests and researchers require a large number of patients and controls to investigate genetic variant combinations and the impact of food on illness. Furthermore, because of this, examining the combined data and determining how multiple genes and nutrients interact is challenging. The following areas face severe challenges:

- –utilizing surrogate/intermediate endpoints;
- –issuing premature health claims;
- –underestimating the financial stakes;
- –misjudgment of moral and legal ramifications

Figure 6.3: Biohacking.

Source: https://www.transparentlabs.com/blogs/all/what-is-biohacking

6.1 PRELIMINARY OUTCOMES

Surrogate endpoints must be implemented. At the outset of a clinical investigation, we seek measurable and accessible criteria that convey a sense of the direction of the study issue. In both earlier situations, the examination of intima-media thickness as a proxy for atherosclerosis is reasonable, understandable, and substantiated by excellent evidence of its connection to more concrete outcomes (Dimitrov et al., 2016). In the second example, the link between polyunsaturated fatty acid consumption and blood levels of HDL cholesterol is a lipid marker connected with coronary heart disease mortality and morbidity.

Surrogates, on the other hand, could offer erroneous and misleading findings and are of little benefit to patients. Endpoints offer a preliminary approximation of the condition we are attempting to prevent, but they must finally be backed by more accurate analyses. The history of clinical pharmacology is littered with stories of drugs that were proven to be beneficial in large-scale trials on an intermediate result, only to inflict damage rather than benefit the same disease process that was known to be connected to the intermediate outcome examined.

Figure 6.4: Nutrigenetics lifestyle diseases.

Source: https://www.firsteatright.com/nutrigenetics-and-nutrigenomics-research/genes-will-rule-what-goes-into-your-plate-in-near-future

Hormone replacement therapy, for instance, was discovered to ameliorate the course of atherosclerosis while boosting cardiovascular mortality due to an unexpectedly considerable extra risk of thrombosis. How far along is current nutrigenetic/nutrigenomic research in determining the ultimate health consequences of tailored nutrition? I assume such instances do not now exist (Dimitrov, 2011). There must be a way to emphasize the therapeutic value of acting differentially on varied patient groups. It is premature to make health claims before the end of this procedure.

6.2 PREMATURE MEDICAL CLAIMS

As a result of the existing insufficiencies in the body of evidence, the bulk of the other challenges originates from the impulse to rapidly exploit the increasing body of newly found information to make hasty, unjustifiable health claims that are instantaneously tied to financial interests. Industrial interests can help the growth of sound scientific research, but they can also hamper it, leading to its final rejection. It is straightforward to appreciate why makers are prepared to sell their genetic tests although physicians do not know how to employ them (Dall'Asta et al., 2022). Similarly, it is simple to imagine (indeed, to watch) the creation of firms selling recipes claiming to be "healthy" and "based on the latest scientific research" and expecting to capitalize on the popularity of tailored nutrition. This is a big issue that must be handled effectively.

6.3 REGULATORY AND MORAL RAMIFICATIONS

In the realms of genetic testing and nutrigenetics, there are ethical and legal considerations that must be acknowledged and dealt with discretion. These are listed below:

- –the management of genetic data;
- –consent, privacy, familial repercussions, and child testing

Employers and insurers may employ information for reasons other than medical care.

In addition, we must take into account the well-established fact that a proper diet and lifestyle are crucial. Therefore, we should avoid clouding these signals with premature supposition and resist the impulse to create overly high standards. It is also necessary to not be scared (Dang et al., 2014).

People whose genetic test findings show an enhanced risk for an illness that may be curable with a pricey and unproven "tailored" diet. In other words, we should continue to analyze what is presently accessible on shop shelves, where there are recognized superior nutritional alternatives to others, regardless of an individual's genetic makeup (Deb et al., 2014). These include foods derived from organic and free-range animal farming (which not only affect our genes but also involve ethical decisions), low-fat products, products with a low glycemic index, an increased intake of fish, fruits, and vegetables, products containing friendly bacteria, folic acid to prevent neural tube defects and severe cases of hyperhomocysteinemia, vitamins for children and older age groups to combat absolute or relative deficiencies, and vitamins for children and older age groups to combat absolute or relative deficiencies.

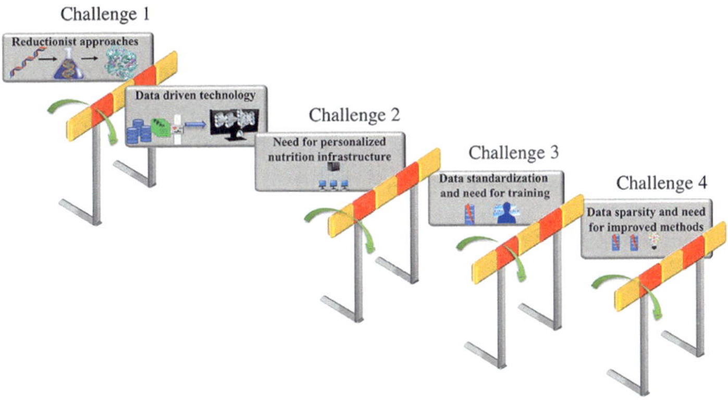

Figure 6.5: Nutrigenomics remains a controversial topic.

Source: Verma M, Hontecillas R, Tubau-Juni N, Abedi V and Bassaganya-Riera J (2018) Challenges in Personalized Nutrition and Health. Front. Nutr. 5:117. doi: 10.3389/fnut.2018.00117

Although there is good evidence for certain gene-diet correlations, others remain equivocal, which is one reason why the adoption of nutrigenomics is controversial. Some medical specialists say that nutrigenomics is not yet suited for therapeutic application and that the majority of the data is obtained from observational studies. Others participate in activities that are not supported by present research. It is erroneous and even hazardous to encourage patients to adjust their diets or lifestyles based on genetics without taking individual clinical biomarkers, dietary and lifestyle choices, and the capacity to adapt into consideration (Connolly & Dawson, 2015). Given the intricacy of the interaction between genes and environmental factors and the hundreds of genes implicated in chronic disease, it may never be realistic to advise a specific diet based purely on genetics.

Other health care professionals are concerned that a nutrigenomic approach will not outperform current dietary counseling procedures. According to one study, a nutrigenomic diet did not help weight loss when compared to a balanced diet. A 2016 meta-analysis indicated that disclosing DNA-based risk assessments has no discernible influence on a diet or physical activity. Another study found that knowing the genotype did not significantly enhance dietary folate intake.

Despite extensive investigation on the relationship between genetics, diet, and health, nutrigenomics is in its infancy. "Nutrigenomics is a pretty young study field, so there is still a lot to discover".

This does not imply that genetics is irrelevant in dietary counseling. Just remember that nutrigenomics will not address every dietary difficulty and that genes are merely one of many aspects that affect health and optimum eating selections (Corella & Ordovas, 2009).

"Genetic testing should not be the major criterion for providing suggestions,". "In our work, we must analyze the patient's lifestyle, health history, health status, personal preferences, cultural identity, change-readiness, and health-related objectives."

Although direct-to-consumer genetic testing for nutritional considerations may appear interesting and practical, there is an extra prospective downside.

The basic problem of these tests, according to Chernus, is that they are not assessed by a skilled doctor. "Skilled practitioners adopt a polygenic approach, which assesses how all genes interact with broader systems in the body." They interpret how these systems interact in the context of one's overall health.

Each gene has several variations: each gene, as far as can be ascertained, has various variants. Different DNA testing businesses may test for different variants, which is why the findings may differ (Chen & Zhao, 2013). One firm may recommend you limit your carbohydrate intake, while another may push you to continue consuming carbs. DNA based diet is a challenging topic.

Simply mentioned, our knowledge and abilities in this area are quite constrained, so stating that we can make particular meal suggestions based on someone's DNA is simply wrong. That doesn't stop the businesses listed above from making these claims.

The tips are quite broad: actually, most of the solutions supplied by these organizations will work for everyone. They are just simple sense.

Furthermore, having a preference for something does not assure that it will occur.

Interactions between genes, vast numbers of genes reacting to a single diet, and contradicting reactions of genes triggered by the same nutrient make nutrigenomics tough (Carlberg, 2019). Aside from the problems of determining the health consequences of nutrients and consequently specific diets and items, it also turns out that establishing the genetic boundary between sickness and health is problematic.

Aside from the challenges and ambiguities of experimental findings, the interaction of nutrigenomics with society is mostly uncertain. How will professional groups such as dieticians, supermarkets, and general practitioners react to the rise of nutrigenomics, its complexity, and uncertainties? What effect will they have on its applications? Will the planned end-users be able to incorporate nutrigenomics goods and services, whatever they may be, into their everyday lives?

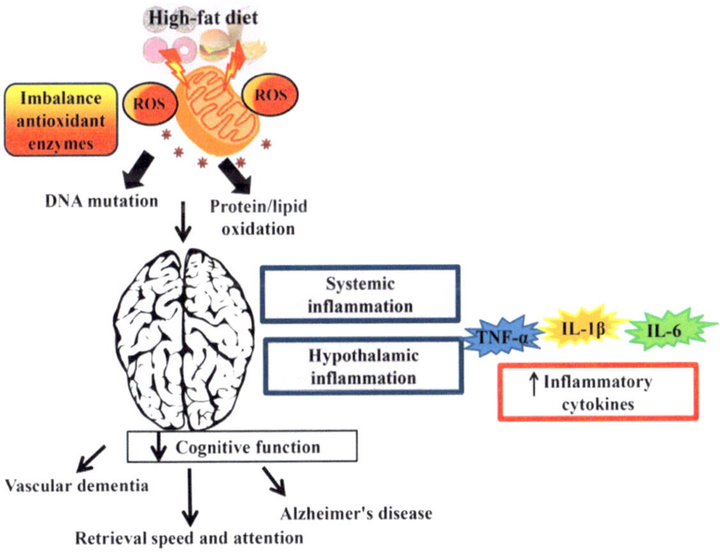

Figure 6.6: Gene responses.

Source: Tan BL, Norhaizan ME. Effect of High-Fat Diets on Oxidative Stress, Cellular Inflammatory Response and Cognitive Function. Nutrients. 2019; 11(11):2579. https://doi.org/10.3390/nu11112579

Nutrigenomics is consequently surrounded by internal and external uncertainty, and end-users would be sensible to better prepare themselves for a genomics future with basic uncertainties rather than expecting the uncertainties to dissipate in the long run. The problem of genetic uncertainty has been addressed before, for example, from the views of policy science management and scientific theory (Bragazzi, 2013). Here, we will focus on the effect of nutrigenomics uncertainties on citizens/consumers by analyzing the application of an ethical stance that focuses on how citizens/consumers cope with basic ambiguity.

Many ethical approaches to genetics or biotechnology, in general, have taken genomics' move toward certainty for granted. They perceive genomics' uncertainties to be temporary and examine norms for managing or even constraining the projected outputs of this research without diving into the science's subtleties.

Uncertainties emerge not simply from contradictory findings, but also from more complicated factors: Cognitive uncertainties, or uncertainty emerging from (social) scientific findings, are associated with moral uncertainties, or doubts that people experience while considering what is the correct thing to do in a particular scenario. Second, we will analyze these doubts in the context of scientific and postmodern concepts, as well as how they link to other movements. Third, we will suggest uncertainty ethics that empowers citizens/consumers to live with more and less basic scientific questions. Finally, we will examine how to cope with ongoing genomics problems.

6.4 UNCERTAINTIES CONCERNING NUTRIGENOMICS, ITS SERVICES, AND COMMODITIES

Nutrigenomics will most likely, and hopefully, not provide serious risks such as those addressed by Frankenstein trajectories; nonetheless, patients and users of nutrigenomics encounter different uncertainties: uncertainties abounding! The bulk of these uncertainties is not addressed by the usual idea of risk or the phrase "incomplete information." Most uncertainties may be grasped, but assessing uncertainties is a unique activity that is not based on new information but, as we shall see later, on ethical techniques, norms, and principles that are now being established (Castro et al., 2006).

A non-exhaustive list of uncertainties that must be addressed in the ethical evaluation of the dynamics of nutrigenomics services and commodities encompasses at least five types, with the first three primarily cognitive and the final two largely social and normative. First, the experimental, internal uncertainties that we said earlier play a role: genes could have diverse functions in interaction with their environment, and the cause-effect connections are primarily unknown till now. Second, it is unknown how frequently polymorphisms would appear in personalized dietary advice depending on personal DNA. Will every individual be prone to a given condition, or will it merely apply to a subset of the population? Third, risk analysis generates uncertainty. Will it be able to measure the probability of having a certain condition with a specified lifestyle? How far can false positives and false negatives be ruled out? How difficult would it be to discern between the inherited and lifestyle components? How significant will these dangers be? Fourth, it is unsure to what degree danger indicators like obesity, diabetes, and other vulnerabilities urge persons to live healthier lives and hence aid to prevent these illnesses (Bionaz et al., 2015). How will the actors deal with these risks? Fifth, it is unknown how nutrigenomics goods and services are generated and consumed. Will it be conceivable to build more effective health-improving products? Is nutrigenomics going to be a commercial tool?

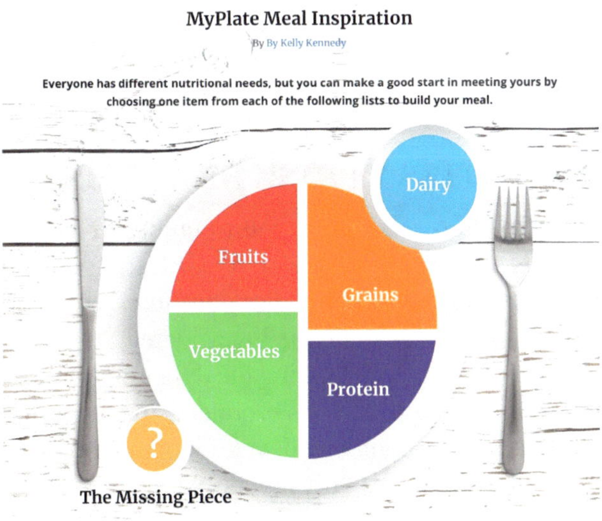

Figure 6.7: Healthy nutrition.

Source: https://www.everydayhealth.com/diet-nutrition/myplate-the-ultimate-guide-to-healthy-eating/

6.5 THE EXPERIMENT'S UNCERTAINTIES

The fundamental uncertainties of the nutrigenomics experiment are particularly impressive. In addition to the more "normal" difficulties of testing, there are additional genomics-specific concerns, such as a lack of data homogeneity and challenges establishing causal linkages based on solid scientific evidence. In contrast to the previous paradigm of one gene-one protein, genes now encode a spectrum of transcripts and proteins (genes could have multiple roles in connection with their environment). They encode proteins that execute a wide range of activities, including those that are diametrically opposing. The phenomenon of moonlighting, gene-splicing, or gene-sharing allows proteins to behave in multiple manners, complicating the route from genotype to phenotype. We don't know how far we'll get in defining whether minerals, foods or even diets are healthy or bad (Brennan & de Roos, 2021). Furthermore, many consequences identified in European populations may not be relevant to other people. Because we don't know what the validity of nutrition and DNA is, experimental uncertainty may be viewed as cognitive uncertainty. This type of uncertainty may have an effect on moral issues around tests that detect early-stage lifestyle-related illnesses or sensitivity to lifestyle-related diseases. It also raises problems regarding the functional food industry's claims. As a result, cognitive ambiguity may affect moral doubt. It may, for example, generate ethical concerns with diagnostic treatments or DNA testing, as well as judgments made for others.

6.6 UNCERTAINTIES IN RISK ANALYSIS

The second sort of cognitive doubt is tied to tailored risk analysis and the subtleties of diet DNA interactions (or diagnostic testing). Risk analysis may be carried out in two ways. To begin, the consumer's susceptibilities as a result of certain polymorphisms may be investigated. A second technique is applied to uncover biomarkers. Following the dynamics of these biomarkers, such as identifying mRNA and proteins produced after a western diet, it may be conceivable to recognize someone developing lifestyle-related disorders at an early stage when they are still reversible. These biomarkers do not yet exist, and their clinical validity and utility must be established. Both kinds of risk analysis pose questions concerning the unknowns (Braicu et al., 2017). To begin, it is unknown how frequently polymorphisms will show in risk estimations based on personal DNA. What are the dangers related to various polymorphisms? Will a certain sickness harm everyone, or will it merely afflict a tiny portion of the population?

Second, how accurate will nutrient susceptibility risk assessments be? Is it realistic to forecast the possibility of developing a given condition based on one's lifestyle? Will it be hard to discriminate between inherited and environmental factors? What does "total risk calculation" include, and how can you decide between a 65 percent risk for one vulnerability and a 45 percent risk for another? One apparent example that genetic testing may be used to predict susceptibility to certain diet-related illnesses is the present obesity research. Jeffrey Friedman, a pioneer in obesity research, contends that inherited factors account for the bulk of obesity variance. He admits the relevance of genetic characteristics but underlines the importance of other factors like energy expenditure from physical activity and effects throughout the uterine period (Braicu et al., 2017). Despite these inconsistencies, extensive research is being undertaken to find markers that may be analyzed to predict the risk of obesity and the illnesses that accompany it. Three variations in the gene that codes for enpp1 have recently been revealed as risk factors for obesity and diabetes. According to BBC News, this "maybe" the first step toward producing a DNA test for obesity and diabetes.

Furthermore, because many health vulnerabilities do not grow to full-fledged illnesses, classifying them as "pre diseases" is premature. Minor disorders that are cured by the body are fairly widespread in everyday life. It's worth mentioning that large-scale population studies repeatedly reveal that many individuals have suffered strokes or brain injury without realizing it and that if there are no following difficulties, there's a significant probability that a "pre-disease" will not develop into a disease due to the body's normal responses. Finally, the organizations that are presently offering these tests on the market (via the internet or elsewhere) are mainly untrustworthy. According to a US Government Accountability Office report, these businesses often defraud their clients.

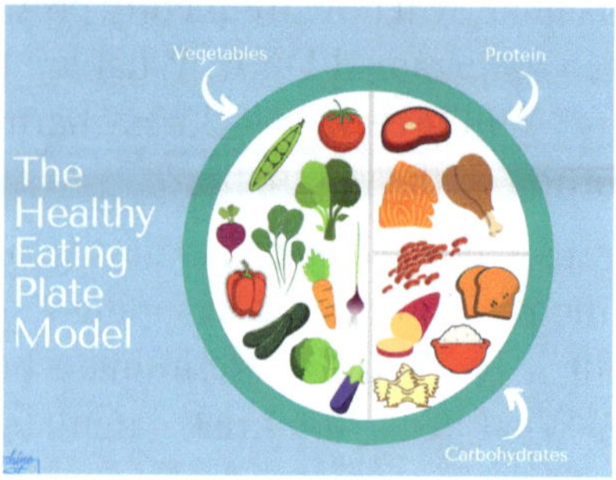

Figure 6.8: Healthy eating concept.

Source: https://sunshinecoastdietetics.com.au/balanced-plate-model/

On the subject of individualized nutrition advice, there is a lot of ambiguity.

Precautionary actions and dietary advice based on risk assessments have also been a subject of worry. Individuals at risk of lifestyle-related illnesses can benefit from more precise advice owing to biomarker and DNA tests. Individualized pharmaceutical and nutrition suggestions might become a reality if genetics is shown to be true (Wiwanitkit, 2012). Adopting a healthy lifestyle and consuming particular foods can minimize risk, but to what extent? As a result of these ethical difficulties, testing, nutritional advice, and the functional food industry are all in danger of being imperiled. If you don't know how to prevent a major peril, can you morally inform someone about it?

A lot of the time, preventative interventions based on predictive testing do not work. According to the study, women who are "at-risk" for osteoporosis should not be offered prophylactic medicines. Is there a policy in place to protect the performers from these threats? Will the findings of the test be communicated to members of the consultee's family who may be in jeopardy? Is it time to give this project over to someone else? What significance do they place on these hazards and partially effective strategies of prevention?

It is feasible to adopt a diet to an individual's DNA profile. They might be used to assess if a person is at an early, reversible stage of lifestyle-related diseases. The examination of a person's diet can help them prevent certain lifestyle-related diseases, such as diabetes and obesity. Clinical validation and utility must be demonstrated for these biomarkers, which were once easily available but are no longer.

6.7 ANXIETIES EXIST IN BOTH THE SOCIAL AND THE MORAL DOMAINS.

The reactions of stakeholders to the awareness of unknown risks and foggy direction are intimately tied to the fourth category of uncertainty: social and moral ambiguity. First and foremost, we must underline that governments, insurance companies, dieticians, doctors in general, food manufacturers, and merchants are unsure about their reactions. Are there any controls in place to guarantee that customers' privacy is safeguarded while still offering better nutritional guidance? The question is whether the bio-data banks will be exposed to the general public, or will be utilized just for certain purposes. What are the implications of telecare and ambient intelligence for service delivery? The second challenge is that it's hard to tell how customers will react to diverse points of view. Is it going to be a good moment for them? Using obesity and diabetes risk indicators to encourage healthy lives may or may not be beneficial in preventing the diseases they point to (Alagawany et al., 2022). Even geneticists recognize that even if a person's DNA reveals that they are in danger, their behavior is unlikely to alter. Individuals' health practices are impacted by a myriad of linked elements. The majority of health behavior models claim that a person's ultimate health behavior is not primarily influenced by their fear of prospective threats. To be accepted in a group, one must agree with the statement being made by others. Because cultural standards on how to actively play a big part in this social influence, cognitive uncertainty and moral doubt are related. The use of either dietary counseling or DNA testing to establish a person's health raises an ethical challenge.

Due to the item's unpredictable nature, nutrigenomics products and services for general usage, such as food, are yet under development and it is not obvious how they will be employed. Genetic testing for lifestyle and nutritional guidance, as well as individualized meals, may be on the way. Lactose intolerance patients and anyone with food-related health disorders may gain from the findings of this study. In addition, it tries to design meals that boost cerebral, musical, and creative talents (enhancement). Their objective is to produce food items that can meet a wide range of dietary demands in the future.

Are there items that can improve health, prevent disease, and boost performance? Nutrigenomics may, on the other hand, largely be employed for profit. Profiting from health-related products and services will always be a concern as long as people care about their well-being. When examining whether nutrigenomics may be utilized to generate more healthy and effective goods, businesses are in charge (Wittwer et al., 2011). Genomic research might become a craze if it is employed in the corporate sector. People will no longer be in charge of the items they buy. For the fifth ambiguity, moral issues may be traced back to the foundations of the issue.

There is a dispute among philosophers of science and social philosophy on the definition of uncertainty. Modern science and social philosophy have evolved conceptions that confront scientific and technical uncertainties, both on an epistemological and moral level, without expecting them to evaporate as knowledge develops.

According to classical epistemology, knowledge and certainty are inseparable ideas. You can only claim to know something if you believe you do. This suggests that you must know all essential facts in order to be assured of something's validity. Popper, for example, contends that scientific efforts are predicated on presumption and uncertainty, which demands a continual re-evaluation, organized critique, and the formulation and refining of new theories to meet them. It was created by American pragmatists to explain scientific institutions' hunt for uncertainty, which they labeled "fallibilism". Science believes that it will someday be feasible to eliminate wrongly reported findings, thus there is space for uncertainty. Personal knowledge and certainty—along with the related authority—are crucial to Popper's theory of ethics.

Before modernity, religious and philosophical traditions, as well as political ideologies, imposed pre-modern ethical frameworks that cleared moral ambiguity. As a result, humans were unable to exercise free will. You were evicted and punished if you didn't follow the same social standards. In today's culture, uncertainty is no longer a given because of the reluctance of many individuals to accept it. Action plans and criteria are available to individuals, but the impact of their efforts is often unpredictable (Van Breda et al., 2015). People in a single area cannot be expected to adhere to the same set of rules and make the same judgments because of the expanding effect of global networks. Expected diversification may be noticed. As a result, individuals are continuously on the hunt for fresh, robust pillars around which to build their own lives.

Beck's groundbreaking social theory on the risk of social and scientific and technical uncertainties is integrally related to Bauman's work on socially-induced uncertainty As a result of the analytical and reductionist nature of the current study, he contends that postmodern (or late modern) civilizations are perpetually faced with new hurdles. There is no need to worry about anything other than your capacity to "remove" non-hacking components from laboratory scientific activity. The relevance of the issue, the researchers' aims, and their own personal interests all influence the selection of variables and the interactions between them that scientists must explore in their laboratory work. One of the challenges with closed systems is that they can never be totally closed, hence taking laboratory data outside of the laboratory exposes one to incomplete information, which is a source of uncertainty (Su & Lu, 2012). For every problem that is identified in the real world, there is a need for more study to find a solution, culminating in extra undesired repercussions. Beck contends that the hunt for ambiguous knowledge, the persistent promise of revelation, the deconstruction of facts, and the continuous revealing of normative inequities constitute our modern civilization.

When presented with so many unknowns, can a person bear the discomfort? There is a limit to how much uncertainty humans can take. Uncertainty in daily life is minimized by the desire for social and relational stability. Uncertainty may be endured to some extent since society gives an anchor of stability. How do people come to accept certain realities as factually correct? How do you manage uncertainty? Mary Douglas ponders the topic, "Do we establish institutions to defend our most treasured ideas?" Emphasis must be made on the significance of an undeniable fundamental conviction that elicits emotions such as "this guy is insane." To him, "the game of questioning implies confidence" is the origin of the human condition. A foundation, i.e., a dependable framework, is sought by a doubt. As long as we are so confident of essential truths, such as the existence of my hand and the fact that I was not on the moon that day, we can only conceive uncertainties.

It's debated whether or not cultural conditions may operate as a mooring point for individual anxieties. Even the strongest social ties can become weaker or stronger over time. People have been more inclined to create new alliances and take on greater risks at particular moments throughout history. Using a unique method, this second-generation American pragmatist investigates the relationship between certainty and ambiguity in his work. In his perspective, there is a great yearning to feel confident in one's own skin. When it comes to regulating ethical behavior, however, Dewey feels that certainty comes at a cost: the loss of discussion and social standards like taboos. One's intellectual growth and cognitive skills are badly damaged by this craving for certainty and the lack of stimulation it demands. People have devised all kinds of things to offer the impression of certainty in a risky and dangerous situation (Afman & Müller, 2006). Many taboos and harsh judgments are retained despite the fact that education has made it feasible for humans to research and live with uncertainty more peacefully. People will continue to seek assurance as long as the world is filled with uncertainty and peril. Individuals may feel more at ease and susceptible to confusing sensations if there is a sense of confidence in the setting. Individuals may pursue a "proud to be uncertain" balance owing to the effect on social and personal confidence.

6.8 THREE THINGS CAN HAPPEN AS A RESULT OF SCIENTIFIC UNCERTAINTY

Genomics, given the present level of scientific and cultural thought, doesn't eradicate all doubts, but rather increases the number of them. Ethically, the five forms of doubt (or, rather, their basic persistence) that we have characterized, are in conflict with the most commonly accepted principles of dominant ethics. Uncertainty isn't addressed in conventional ethics as it believes that principles and standards may be adhered to under constant and continuous settings. It is based on the notion of a long-lived, stable universe. Responsibility informed decision-making, and consumer control is all well-known moral ideas. The knowledge that may be used as a foundation for decision-making must be available at all times to guarantee informed consent (Singh & Sharma, 2019). A person's acts, as well as the outcomes and repercussions of those activities, must be tied together. Consistent personality and social environment are also required for successful consumer control. We must bring new collaborative ways of thinking and decision-making that all sides must adjust their opinions and interests to the continuous discussion in order for the community to gain. We can conclude that protecting one's own interests against threats from government and professions is a hollow ethical precept. Participation in contemporary talks about nutrigenomics concerns demands a code of ethics.

Consumers and individuals alike are seeking a code of behavior for coping with uncertainty. Conversations, debates, and the exchange of tales and life narratives are common activities throughout the selection process, but there are no consistent procedures or rules. You must understand the connections and contrasts between your life and the lives of others touched. Putnam's contrast between common sense doubt and philosophical doubt, which refers to the possibility of situations that are more or less obvious, may offer some ethical justification.

The distinction between common sense skepticism and philosophical doubt is crucial. Humans lack the capacity to cease believing in anything. Criticism is inadequate to explain widespread distrust. The fact that we commit errors is not sufficient grounds to change our opinions.

In daily life, there are various forms of uncertainty, each of which is either more or less accurate. In this context, "informed choice" is not a relevant ethical term as it does not address how to choose between more certain and less certain options or the incorporation of known health features into one's diet and personal health (Sikalidis, 2019). We require ethical categories, such as consultation and deliberation, that assist consumers in making these decisions, as well as learning processes that allow consumers to share their personal stories (narrative input) and be assisted in becoming more resilient to health problems that are in their best interests. This ethic encourages consumers and professionals to connect in order to bridge the gap between food production and consumption, new technology, and healthy eating habits.

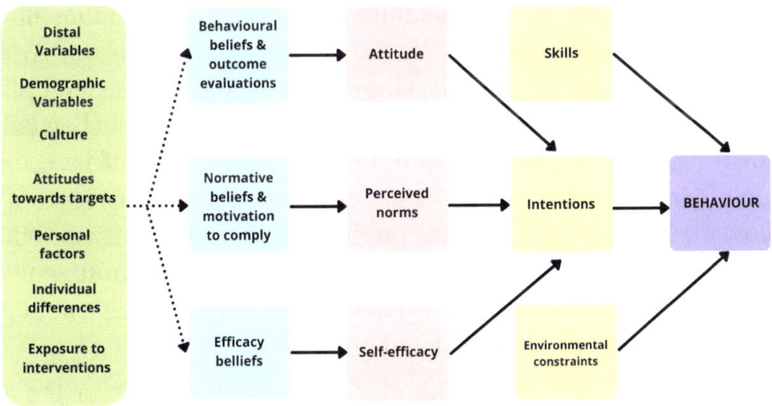

Figure 6.9: Ethical issues for human nutrition.

Source: Dix CF, Brennan L, Reid M, McCaffrey TA, Molenaar A, Barklamb A, Chin S, Truby H. Nutrition Meets Social Marketing: Targeting Health Promotion Campaigns to Young Adults Using the Living and Eating for Health Segments. Nutrients. 2021; 13(9):3151. https://doi.org/10.3390/nu13093151

To guarantee that all opinions and points of view on a particular issue are heard, remarked on, and integrated into the meaning frameworks that finally underpin political decisions made by elected officials, discussions should be done. The purpose of debates is to include the community in the process of self-discovery, not to gain consensus or make choices. These scenarios are a fantastic starting point for examining how a specific action may impact the future.

Consider Dewey's idea of "happy with uncertainties" if you're searching for a strategy to balance your personal and social life. Is there anything you can do to obtain happiness despite adversity? We have returned with some philosophical wisdom. A well-balanced assortment of choices is typically the most successful strategy. Managing uncertainty is as simple as dismissing the unpleasant signals that alert us that we are in danger and continuing as normal. Additionally, it is possible to understand one's sensitivity and take every precaution to prevent the disease's early stages.

The notion of being informed you're in danger with no viable solutions has been mentioned earlier. Strong emotions that distract from one's physical condition could be exploited to handle uncertainty. Spinoza's ethical strategy is to lessen anxiety through other feelings, like joy while eating wonderful food, instead of worrying about one's own health every time one consumes something. Positive thinking can be used to rationalize and sublimate worry. This is verified by the Nicomachean Ethics of Aristotle. According to him, short-term aches and pains should take second place to long-term health advantages (Riscuta, 2016). Even when challenged with one's own reservations, Dewey would highlight the need of investigating them. In other words, it is equivalent to undertaking a (moral) investigation to discover whether new lifestyles, products, and preferences are ethically desirable. Consequently, you are stepping outside your comfort zone while being miserable. The connection between a person and their environment changes the balance between trust and uncertainty.

In discussions about the health effects of specific foods or the ability to prevent or delay the onset of a disease, cognitive and normative claims interact. What claims should you believe and which should you disregard? How much can we rely on privately owned small businesses? In light of these challenges, we must decide who we can trust. It is obvious that a trusting relationship between nutrigenomics and its producers and end-users is required.

Genuine discussions and conversations can reveal new collaboration opportunities, such as co-managing technological options or selecting morally preferable solutions from a wide range of options. To make a living in today's society, generics-based products and services, such as high-tech foods and medicines, require the use of procedural and substantive elements of argumentation. All participants must be able to participate in deliberations and agenda-setting. They seek normative guidance in areas such as living and cooperating values (Banerjee et al., 2015). High-tech devices should be used in daily life only if they are integrated into the social interactions of peaceful and purposeful people. Scenarios can be extremely useful in predicting the future.

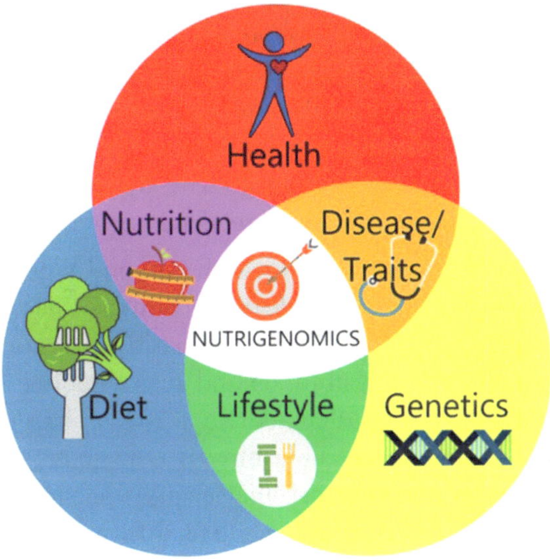

Figure 6.10: Healthy living.

Source: https://goqii.com/blog/nutrigenomics-you/

There are a few more reasons why we should be concerned about our health and bodies. But how much money do we have? This is a difficult problem to solve because the link between diet and genes is only one piece of the puzzle. These findings also suggest that traditional notions of what constitutes food versus pharmaceuticals, as well as healthy living versus disease, should be reconsidered. Consumers are finding it increasingly difficult to distinguish between food and medicine. The Hippocratic adage "let food be your medicine" demonstrates that the line between the two has always been hazy and finicky. The playing field has shifted as a result of recent advances in nutrient-nutrient DNA interaction-based prevention

(Riscuta & Dumitrescu, 2012). He contrasts the dietetic method of accepting one's fate with the therapeutic method of avoiding the illness at all costs. We should not be afraid of contracting a disease if we are confident in our own health, according to the dietetic way of life.

For the sake of argument, let us assume that food has multiple effects on the body, that the line between health and disease is hazy, and that genes can respond to their environment in a variety of ways (both positive and negative, for humans). However, the alleged polymorphisms may present opportunities and challenges in medical and non-medical fields, as well as vulnerabilities. As a result, a person's choice of the health-care system generates a dynamic social environment that influences gene expression. The moral sensitivity can be revealed through scenario creation, which involves imagining these distinct situations with diverse genetic connections. There are ethical concerns about which healthcare systems should be used to generate genomic breakthroughs, products, and services.

Some individuals or groups will use the second type of uncertainty to deny the existence of these risks, while others will use this issue because they believe it will be advantageous. Several scenarios in genomics governance may elicit different moral sensibilities and make deliberative participants aware of morally relevant challenges (Ray, 2014). The use of genomics as a marketing tool will be possible only if consumers and governments collaborate to develop a framework that ensures accurate and transparent health claims are made. This will eliminate the fifth and final source of doubt about the use of genomics.

Imagining future stories or scenarios with various genetic outcomes can help to alleviate some of the ambiguity. Fantastical worlds created by artists may also be valuable. Patients, consumers, and citizens must all play a role in determining the appropriate balance of certainty and uncertainty. To put it another way, it is necessary to separate the significant uncertainties from those that appear insignificant or irrelevant. John Dewey, a pragmatist, referred to their discussions as "dramatic rehearsals."

6.9 CONSEQUENCES FOR BOTH CUSTOMERS AND THEIR SOCIAL ENVIRONMENT

However, because the consequences for consumers cannot be considered in isolation, we will also examine the key agencies operating in this context in this subsection, albeit only in terms of their relationship with consumers.

Consumers can choose to avoid the genetics supply chain entirely. However, it is difficult to remove nutrigenomics and stop using it to treat obesity. Given the wide disparity between the dietary options available and the lifestyles and health profiles of the people who consume them, this is not the best course of action at a time when substantial resources have been dedicated to genetic research on the intellectual, economic, and political levels. The question is how to consume in the new situation of a long life dominated by passivity. What can we do to restore harmony between our bodies and the rest of the world? Nutrigenomics appears to be required for consumers/citizens to fine-tune their food to their new lifestyles. It's difficult to know what foods are healthy in today's mostly sedentary world (Qadeer & Ashhar, 2014). Traditional food recipes are associated with a physically demanding lifestyle, whereas modern lifestyles are more sedentary. Furthermore, during the selection process, consumers require assistance distinguishing between major and minor uncertainties, as well as products and services. To begin with, new educational and dialogue institutions would need to be established in order to bridge the gap between production and consumption. Third, consumers can request that the government establish health or quasi-health claim standards that apply to the entire product rather than isolated features (such as "light"). Customers can now ask the government for assistance in determining the origin and quality of goods in order to better participate in and coordinate public discourse. According to a recent Eurobarometer survey, consumer groups in the food industry have the highest level of credibility (33 percent).

Third parties play a role when it comes to customers. Government agencies, scientists and technologists, and genomics companies may all encourage customers to live with uncertainty in different ways. The government can democratically organize nutrigenomics research, with an emphasis on the prevention of common illnesses and conditions rather than personalized nutrition. Nutrigenomics has a place in public health, particularly in broad risk profiles. Governments should exercise control over three areas: food, medicine, and the newly emerging grey zone, improve research ethics and medical testing of healthy foods, and only allow low-cost health foods... Nutrition policy should prioritize encouraging people to eat with joy in order to foster formal relationships. As health should be a secondary goal of eating, consumer groups should have a say not only in downstream nutrigenomics (which deals with finished products) but also in upstream nutrigenomics research agendas (Pauletto et al., 2020). Because consumers are susceptible to expectations and hypes, it is critical to establish independent gatekeepers to protect the independence of testing, promoting, communicating, and providing services in nutrigenomics. The best way to maintain public trust in the food and health industries is to establish and maintain independent organizations that set and monitor standards for health and enhancement claims.

This implies that genomics researchers who assist consumers in making informed decisions about their own health care, like medical researchers, should pay more attention to research ethics and professional ethics. Unreproducible and/or invalid research should not be published. Because of the complexities of nutrigenomics research, research results on minor details should be disclosed only in conjunction with the larger context, and this should be prohibited outside of scientific circles. Because of the complexity of the interactions between the genome, genetic profiles, and environmental factors such as diet and lifestyle, it is extremely difficult to replicate experiments in other locations. Food and lifestyles are so dissimilar that they can't be left out of the equation. The primary focus of genomics is the relationship between the expression of a single gene and the genome as a whole, rather than a single gene in isolation (Peña-Romero et al., 2018). A variety of factors influence food preferences, including genetics, lifestyle, and ethnic diversity. To prove or disprove causal links, expensive and time-consuming research methods are required. Critical nutrigenomics experts have concluded that consumers should be warned of the high risk of fraudulent claims and unrealistic expectations. Finally, there should be no conflicts of interest. Most scientists agree that they should not avoid difficult subjects, but rather work to better understand the public's concerns (Piroddi et al., 2017). Because there is little evidence to support the development of personalized nutrition, a public debate about the appropriate proportion of public health research to research on personalized nutrition is urgently needed. It is critical to maintaining order in common ailment research, such as that on low-cost and readily available remedies and foods. Healthy food access for all must remain a top priority.

CHAPTER 7

NUTRIGENOMICS AND ITS APPLICATIONS

CONTENTS

In nutrigenomics, gene expression and metabolic pathways are studied concerning the food. Recent interest in nutrigenomics has been spurred by its ability to cure chronic diseases. The importance of nutrigenomics in human health and animal husbandry, particularly swine and poultry, was brought to light in this study. An important factor in human health is the expression of specific nutrition-related genes, and this knowledge is useful for designing personalized diets. The human genome project discovered many polymorphism genes associated with metabolic functions. Lactose intolerance and phenylketonuria are the results of a genetic mutation in gene (Pal, 2018). It is imperative to maximize feed efficiency in the swine and poultry sectors due to the high cost of raw materials. Pigs and chickens' growth and health have been studied extensively using a variety of dietary components, but more precision and knowledge are needed. Nutrigenomics research is still in its infancy, but it has the potential to revolutionize the food industry and individual development. More sustainable fisheries, higher productivity, and more aquaculture operations will be possible as we get a greater understanding of these systems.

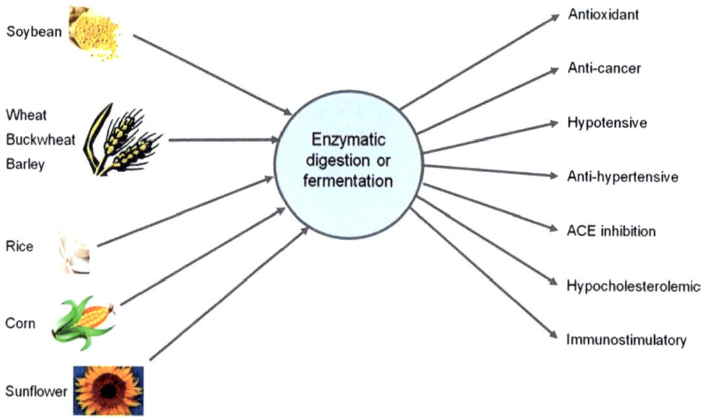

Figure 7.1: Importance of nutrigenomics in soy protein.

Source: Kussmann M and Van Bladeren PJ (2011) The extended nutrigenomics – understanding the interplay between the genomes of food, gut microbes, and human host. Front. Gene. ***2****:21. doi: 10.3389/fgene.2011.00021*

7.1 INTRODUCTION

It has been shown in recent years that proper nutrition and diet can aid in the treatment of many different diseases, such as cancer and heart disease.

The study of how food affects gene expression is known as nutrigenomics, a term coined by nutrition genomicists. To investigate the significance of diet and diet formulation, scientists need to be familiar with the physiological, biochemical, and metabolic pathways that organisms use to respond to dietary components. For the first time, scientists in the United States have looked at how genes interact with the environment to cause a wide range of human diseases (Neeha & Kinth, 2013). An individual's genetic makeup is thought to influence their response to environmental stimuli in a wide range of ways. It is now possible to study the relationship between diet, metabolic pathways, and gene expression thanks to advances in molecular biology techniques. Individuals' dietary needs are now widely accepted to be different and can lead to a variety of health problems, such as weight gain, high blood pressure, and high blood sugar levels. The lipid profile can be improved with nut consumption, a high-carbohydrate diet rich in protein, green tea, and red wine, and supplements like policosanol and red yeast rice extract.

However, supplements like guggulipid, garlic, chromium, vitamin C, magnesium-pyridoxal phosphate glutamate, tocotrienols, and absorbitol aren't suggested. Another study, on the other hand, found a connection between a poor diet and an increased risk of prostate cancer. Genetic variation in the lactase gene, which causes lactose intolerance, is a common example of nutrigenomics. A lactose-free diet is the best treatment for lactose intolerance symptoms. An example of a metabolic pathway disorder linked to nutrigenomics can be found in the condition known as Phenylketonuria (PKU). Human disease risk can be reduced by following a healthy diet, according to the findings of this study. Multiple research groups are investigating the link between dietary and nutritional factors and gene expression, and metabolic and physiological changes in the body (McNamara, 2010). Our understanding of homeostasis, gene expression, and metabolic pathways will be improved by this research. Nutrigenomics research is essential for human health, as well as for the poultry and pork industries. Consumption of poultry and pork has skyrocketed over the last few years. Nutrigenomics research has the potential to improve feed efficiency in these industries. There is still a lot of work to be done in this area. We will be able to create more sustainable fisheries, increase production, and engage in additional aquaculture activities if we have a better understanding of these mechanisms.

The nutrients and other components of food have a major impact on gene expression and transcription control. It is a huge plan that extends beyond the essential sustenance offered by food. SNPs, or single nucleotide polymorphisms, have been discovered to create changes in gene expression patterns across humans. As a result, the types and quantity of food eaten are crucial to one's health and growth (Benítez et al., 2017). Several nutrigenomics investigations have revealed that diets containing bioactive compounds are favorable to health. Nutrigenomics study provides a worldwide evaluation of gene expression and its repercussions, providing fundamentally new techniques for nutritional research.

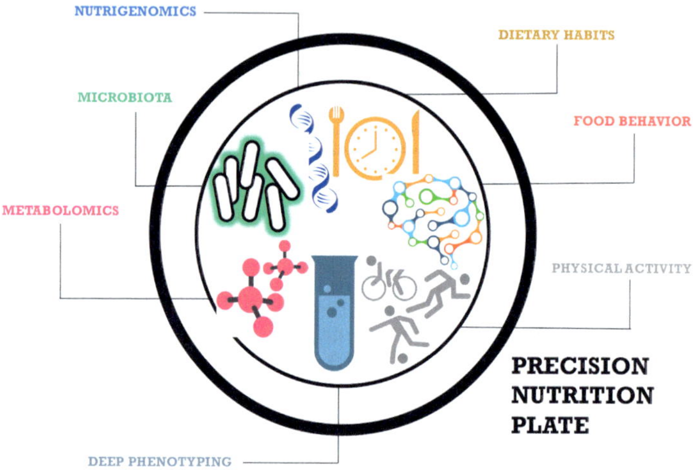

Figure 7.2: Precision Nutrition.

Source: De Toro-Martín J, Arsenault BJ, Després J-P, Vohl M-C. Precision Nutrition: A Review of Personalized Nutritional Approaches for the Prevention and Management of Metabolic Syndrome. Nutrients. 2017; 9(8):913. https://doi.org/10.3390/nu9080913

7.2 NUTRIGENOMIC SCIENCE AND HUMAN HEALTH

Nutrigenomics has lately garnered notoriety owing to its promise for preventing, relieving, or curing chronic sickness and diverse cancers. SNPs make up only 0.1 percent of the variation in human genomes, which are 99.9 percent alike. According to the study, the human genome has around 3.1 million SNPs with high polymorphism, resulting in individual diversity. Several studies have indicated that various macronutrients (protein, carbohydrates), micronutrients (vitamins), and naturally occurring chemicals (flavonoids, coumarins, and carotenoids) can influence gene regulation. This SNP database may be used to find genetic differences in populations (Bragazzi, 2013). Cancer and cardiovascular disease (CVD) are two of the most frequent ailments that dietary modifications can influence. Recognizing dietary risk factors is crucial as it will surely assist to reduce the population's exposure to these variables. A variety of ailments have been connected to dietary components. Here are a few instances of noteworthy situations.

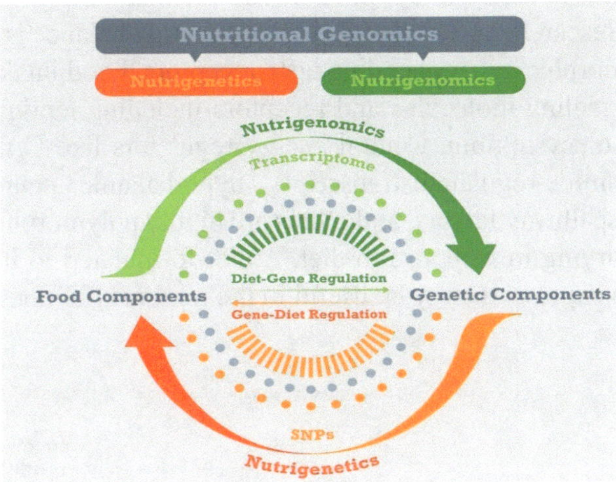

Figure 7.3: DNA & nutritional genomics.

Source: https://www.acecgtnutrigene.com/genomics.htm

Cardiovascular disease: In recent years, people's eating habits have shifted dramatically, resulting in a range of health risks. Cardiovascular disease (CVD) symptoms include obesity, hypertension, and thrombosis . Apolipoprotein A-1, a protein related to high-density lipoprotein (HDL), is synthesized in the liver and intestines . The A-1 HDL complex is critical for cholesterol transport and has been associated with a lower risk of cardiovascular disease. The polymorphic APOA1 gene, which codes for Apo A-1, has been intensively investigated. Variants of this gene have a significant influence on cholesterol transport, which leads to CVD. Polyunsaturated fatty acids (PUFAs) are thought to activate the APOA1 gene (PUFA). CVD in this area may be brought on by changes in lipid metabolism. Studying PUFA's impact on cholesterol transfer in more depth, it was shown that patients with the APOA1 A allele had greater HDL levels as their PUFA intake increased. On the other hand, those with the APOA1 G variant exhibited decreased HDL cholesterol levels (Braicu et al., 2017).

Cardioprotective cuisine was rated higher than heart disease-promoting meals regardless of the presence of the metabolic syndrome. To avoid CVDs, a customized diet is essential.

Weight gain is another complication of a poor diet. Many metabolic syndromes are influenced by both environmental and genetic factors, according to researchers. The body's delicate energy balance is the result of several polymorphic genes encoding taste receptors. Food intake is affected by several signaling molecules and receptors including leptin and insulin, as well as cholecystokinin. Central energy regulators like CART (cocaine and amphetamine-regulated transcript), hypothalamic neuropeptide Y, melanocortin pathway factors, and others are highly polymorphic, with their expression varying in response to diet. A diet formulated to influence and regulate gene expression may be useful in the treatment of obesity.

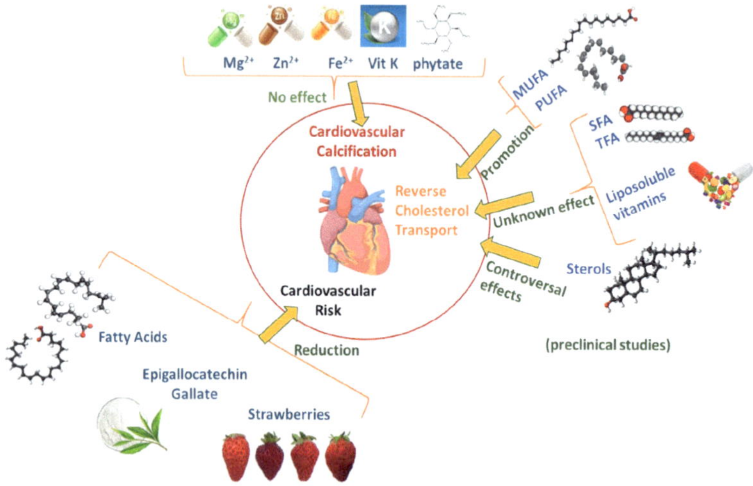

Figure 7.4: Cardiovascular diseases.

Source: Adorni MP, Ferri N. Nutrition Intervention and Cardiovascular Disease. Nutrients. 2022; 14(7):1435. https://doi.org/10.3390/nu14071435

Oncology treatment: Recently developed molecular biological techniques have provided new insights into a wide range of health problems including cancer. There are indeed many options for cancer treatment right now, but they all come with unfavorable risks and effects. It takes a lot more energy for normal cells to multiply at a slower rate than it does for cancer cells to grow. The metabolic pathway has been altered to meet the rising energy demand. Glycolysis and lactic acid fermentation are the primary sources of energy for cancer cells. As a result, cancer cells rely on glucose for growth. The expression of glucose-sensitive genes like pyruvate kinase type M2 and phosphofructokinase1 could be used as an alternative cancer treatment (Mustapa et al., 2020). Nutrients found in fresh produce have been linked to a decreased risk of cancer. Methionine, which aids in DNA mutilation, is derived from folic acid in fruits, which is converted to 5-methyltetrahydrofolate through a series of chemical reactions.

A diet deficient in the B vitamin folic acid has been linked to a higher risk of cancer due to the inhibition of DNA replication. 1,25-dihydroxyvitamin D3 regulates numerous bodily functions, including cell proliferation, through the polymorphic Vitamin D Receptor (VDR). Colorectal cancer can be caused by Fok1 mutations (a polymorphic form of VDR). PTEN, cyclins, p53, and Wnt have been linked to angiogenesis and cell proliferation-related gene expression, which can influence tumorigenesis in numerous studies. According to research, several minerals may play a role in the development of cancer. A few examples include selenium (which boosts enzyme production of glutathione peroxidase and reduces stress), prostacyclins (which guard DNA and lipids against oxidative damage), and zinc (which regulates gene expression and stability). As a result, to minimize health risks, customized diets are required.

Obesity and cardiovascular disease have been linked to hypertension, which is a serious condition. The presence of sodium chloride in the diet has been linked to an increased risk of hypertension. According to the findings of this study, 15% of the population is salt-sensitive, while the other 85% do not react. These two polymorphic genes, AGT and aldosterone synthetase, play critical roles in the control of blood pressure (CYP11B2). Dietary manipulation of these genes could be used to treat high blood pressure (Masotti et al., 2010).

Applicability of nutrigenomics in the animal sector: In recent years, nutrition research acquires a special emphasis due to its extensive utilization in numerous areas of science. It has been established that not only environmental signals but various additional components are associated with animal health. Nutritional genomics is a recent off-shoot of this genetic revolution. It was argued that nutrigenomics is the nexus that links nutrition, health, and genomics in terms of phenotypic effect. For instance, nutrigenomics is prominent research that integrates numerous distinct omics similar to proteomics and transcriptomics. The right synchronization between diet and omics is a key critical regulator in sustaining animal health. Recent research suggested that bioactive material contained in meals alone operate as a transcriptional factor or interact with transcription factor and influence the expression of metabolic genes. Several studies found that vitamin D, A, and fatty acids may stimulate the transcription of genes by interacting with nuclear receptors. The current advancement in molecular biology has provided several ways to explore the association between nutrition and gene. Among them, DNA microarray has proved its promise. In general DNA microarray is a sophisticated technology that tests the expression of genes or mRNAs level at a time (Mathers, 2017). The principle of DNA array is the hybridization of 2 standard DNA sets, which is defined by hydrogen or covalent bonding. The detection of gene expression is done by deploying numerous fluorescent probes which are classified into two types depending on the array: single labeling and multiple labeling.

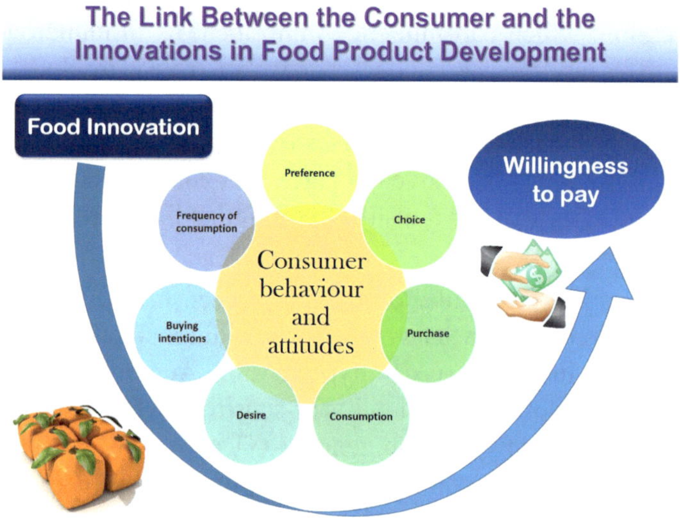

Figure 7.5: Dietary guidance and food product development.

Source: *Guiné RPF, Florença SG, Barroca MJ, Anjos O. The Link between the Consumer and the Innovations in Food Product Development. Foods. 2020; 9(9):1317. https://doi.org/10.3390/foods9091317*

In the pig business, goat farms, and broiler chicken farms, the concept of nutrigenomics is being increasingly used. Increasing globalization and population growth have resulted in a huge increase in the demand for pork across the world. In recent years, biofuel companies have used a large number of cereals to produce biodiesel and ethanol, which has pushed up grain prices. For this reason, nutrigenomics is the only way to reduce the number of raw materials used in animal farms while still increasing profit.

Researchers are eager to improve feed efficiency by experimenting with various feed combinations. Pigs receiving glutamine supplementation had higher levels of gene expression in the stomach, according to a microarray study. Researchers in another study looked at how protein and carbohydrate ratios affected fat synthesis and gene expression in pig muscles (heart fatty acid-binding protein or H-FABP, peroxisome proliferator-activated receptor or PPAR). Increasing intramuscular fat by regulation of a previously known gene was found to improve the quality of pig meat on a low protein diet, according to the researchers (Loor et al., 2015). Bacteria in the intestines play an important function in the nutrition and health of the host. The composition of the microbiota in an animal's digestive system has a direct impact on the animal's metabolism. RS (Resistant Search) is not an enzyme that pigs produce on their own, thus they rely on microbes in their digestive tract to convert it to short-chain fatty acid (SCFA). SLC16A1 and glucagon (GCG) production may be boosted by a diet high in RS, which affects the microbiota of the colon and caecal, according to the hypothesis (GCG). One of the most important factors affecting the quality, safety, and nutritional value of meat is the feeding schedule. Researchers studied the effects of dietary lipids such as beef tallow, soybean oil, olive oil, and coconut oil on the health of pigs using DNA microarray and real-time PCR. CAMP-dependent regulatory, protein kinase type II alpha (PRKAR2A) and the catalytic subunit of protein phosphatase 1, beta isoform (PPP1CB) were upregulated in groups treated with olive oil.

In growing-finishing pigs, the link between dietary fat and gene expression was further shown. Pigs require nutritional protein in addition to lipids for optimal health. Pigs were fed casein and soy protein as sources of protein in an experiment. There were substantial differences in genes such as organic anion transporter polypeptide 2 and calnexin, glutathione-S-transferases, and methionine-Sulfoxide reductases across the two groups of animals given soy protein and casein protein, respectively (Ladeira et al., 2018). Hepatic gene expression of Cytochrome 8b1 (CYP8B1), aldehyde dehydrogenase 2 (Aldh2), and thiosulfate transferase (TST) were all increased in pigs fed *Camelina sativa*. To improve feed efficiency, pigs must express nutrition-related genes. However, there are very few studies on pig nutrition. Growth, production, and meat quality will all benefit from a better knowledge of how this system works.

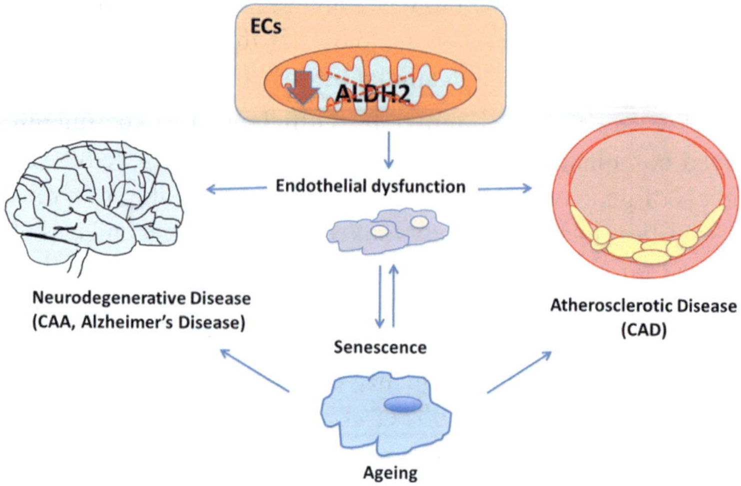

Figure 7.6: Aldehyde dehydrogenase 2 (Aldh2).

Source: Nannelli, Ginevra & Ziche, Marina & Donnini, Sandra & Morbidelli, Lucia. (2020). Endothelial Aldehyde Dehydrogenase 2 as a Target to Maintain Vascular Wellness and Function in Ageing. Biomedicines. 8. 4. 10.3390/biomedicines8010004.

There is a growing demand for broiler chicken and pig meat in the poultry industry. A chicken farm's feed cannot be optimized without the help of nutrigenomics research, which is essential for the animals' health and productivity. Another important aspect of broiler production is that nutrigenomics plays an important role in multiple ways, including (a) correlating nutrition and genetics, (b) epigenetics, (c) increasing feed efficiency, (d) improving health, and (e) increasing meat quality. There has been a slew of studies in the last few years looking at the impact of diet on the neonatal and early stages of life (Kussmann & Affolter, 2009). Adult broilers' body weight and meat quality are both reduced after a 24-hour fast, according to a new study. In addition, it was claimed that chicks' growth and development could be accelerated by feeding them a low-protein diet.

Nutrigenomics research, in particular, will aid the poultry industry in developing more precise feeding plans. Arbor Acres (AA) and Beijing-Young (BY) female broiler strains, according to a recent study, both have a high nicotinic acid content (BJY). Another interesting finding was the presence of both ApoA-I and ApoB mRNA expression in the liver, which was found to correlate with the feed's concentration of nicotinic acid.

Bioactive substances, as well as trace elements and vitamins, are also thought to have health benefits. In broiler chickens, vitamin E acts as a transcriptional regulator of genes involved in lipid oxidation and antioxidant gene expression, reducing stress and improving meat quality. Adding vitamin E to chickens' diets has been shown to improve their ability to fight off bacterial lipopolysaccharide infection. Chickens that eat a diet high in algae have a lower stress level, according to nutrigenomics data. As with vitamin E, the amount of zinc in the feed has a direct effect on the expression of the transporter gene in the intestines of broiler chickens, and this effect is similar to that of zinc (Brennan & de Roos, 2021). However, Selenium is one of the Yuan regulators of gender-specific gene regulation in chickens. Mineral excretion in the feces was reduced while growth was maintained on the Mintrex-Zn/Mn diet. Several other studies have found that different diets affect the gene expression of chickens.

Antibiotics, vaccinations, and other treatments are randomly provided on a poultry farm to deal with disease, which enhances the coast. Diet based tool might be an additional strategy to lower poultry mortality. Till now, it remains a novel field of research. More investigation should be performed to understand the diet-linked genetic network in an animal. In these regards, the nutrigenomics database will be the possible instrument.

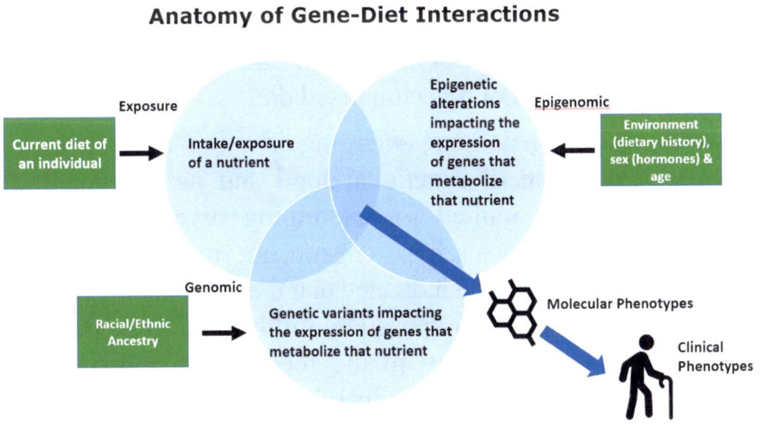

Figure 7.7: Nutrigenomics and internal balance.

Source: Mullins VA, Bresette W, Johnstone L, Hallmark B, Chilton FH. Genomics in Personalized Nutrition: Can You "Eat for Your Genes"? Nutrients. 2020; 12(10):3118. https://doi.org/10.3390/nu12103118

7.3 FUTURE SCOPE OF NUTRIGENOMICS

New tools accessible in current research allow dietitians to investigate genetic background through transcriptomics, proteomics, and metabolomics and develop dietary plans for personalized nutrition. The Nutrigenomics approach may be classed into 3 ways (1) Gene switching, (2) Emphasis on gene-protein interaction, and (3) Influence of food components on gene expression. To examine the association between diets and genes, DNA microarray technology and quantitative real-time Polymerase Chain Reaction (PCR) can be applied. Two-dimensional gel electrophoresis might be an essential approach to assess the impact of each amino acid on protein composition that leads to the safe use of transgenic fish in our diet. The division of biomarker nutrigenomics is a revolutionary new approach to examining the food and the pharmacological characteristics of the food to cure sickness and slow down the process of aging (Kussmann & Van Bladeren, 2011). To attain desired health, genetically particular dietary components should be addressed. In Japan, over 350 items have been designated as food for special health purposes by the Japanese Ministry of Health, Labour, and Welfare. Each item has a specific health claim, such as food for hypertension, high cholesterol, diabetes, etc. based on scientific discoveries. Nutrigenomics is a recent issue and there is a lack of research addressing the nutrition-linked gene expression. More inquiry should be done to examine the genetic polymorphism and nutrition-connected gene expression for the creation of a personalized diet.

In conclusion, nutrigenomics, emerging as a new age of study, allows us the greater expansion of functional food and nutraceutical research. Nutrigenomics is often applied for examining several human ailments including heart-related illnesses, hypertension, obesity, as well as diet-related problems. Gene-diet interaction indicated that the adiponectin gene variation relates to insulin resistance. Further study on the subject of nutrigenomics will surely prove its potentiality in the food sector and the growth of humans. This nutritional study is not just for use in laboratories but for the better development of food for society (Kang, 2012). Thus, it is necessary to take care of public information about these challenges. In field application, the producers require control and assistance to boost safety dependability and health advantages on the things. To cure food-connected diseases and enhance health, we can construct Food-Based Dietary Guidelines (FBDG) (. In this post-genetic era, the genes we inherit cannot be altered but we may modify their response to their surroundings by employing nutrigenomics to accomplish desired consequences.

CHAPTER 8

NUTRIGENOMIC LINKS AND CHRONIC DISEASES

CONTENTS

The existence of a certain gene or mutation frequently indicates a predisposition to a disease process. The interaction of an individual's DNA, environment, and behavior govern their capacity to manifest their genetic potential. As a result of this improved understanding of genetics, transdisciplinary approaches to health and disease research are already spreading. Nutrigenomics, for example, combines genetic studies with nutrition and, if possible, other lifestyle factors such as cigarette smoking and alcohol use. While genes play an important part in determining function, the diet has a substantial influence on the amount of expression of specific genes and, therefore, whether or not individuals achieve their genetic potential. As a result, the term "nutrigenomics" was coined to describe a study into how diet affects a person's genetic make-up. This concept has lately been expanded to include dietary components that preserve DNA (Irimie et al., 2019). The study of the genome, proteome, and metabolome in connection to dietary components is known as nutritional genomics (the sum of all metabolites). Researchers know that similar to pharmacogenomics, where a drug has different effects on different populations, some people may respond positively to specific dietary treatments, while others will be resistive, and still, others will be severely influenced by these therapies.

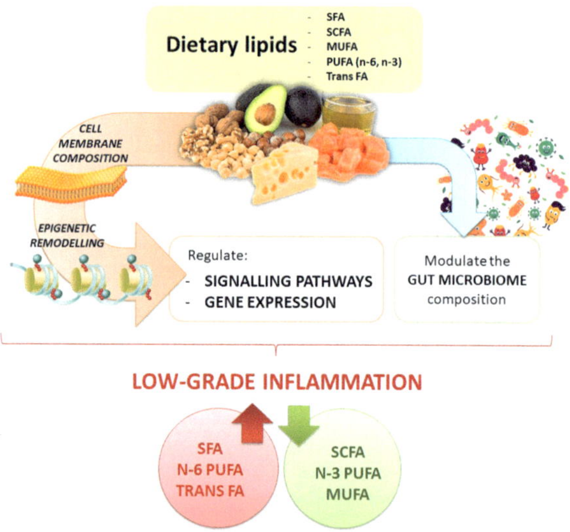

Figure 8.1: Nutrigenomics and gut microbes.

Source: Bordoni L, Petracci I, Zhao F, Min W, Pierella E, Assmann TS, Martinez JA, Gabbianelli R. Nutrigenomics of Dietary Lipids. Antioxidants. 2021; 10(7):994. https://doi.org/10.3390/antiox10070994

8.1 THE FOCUS ON POLYMORPHISMS

Numerous research in people, animals, and cell cultures have revealed that macronutrients (such as fatty acids and proteins) alter metabolic responses. These studies also reveal that vitamins and naturally occurring bio-reactive chemicals (such as flavonoids and carotenoids, as well as zoochemicals like eicosapentaenoic acid and docosahexaenoic acid) affect metabolic reactions. Transcription factors like resveratrol can alter gene expression directly via dietary biochemicals like genistein. Choline indirectly impacts gene expression by changing signal transduction pathways and chromatin structure. According to an increasing body of research, genomic instability may be an early warning indication of nutritional inadequacy even when no clear genotoxicant exposure is present (Hurlimann et al., 2014).

Single nucleotide polymorphisms (SNPs) account for more than 90% of human genetic diversity, which has piqued the interest of the nutrigenomic industry. SNPs that change the activation of "housekeeping genes," which are essential for cell maintenance, have been linked to an increased risk of illness. Nutritional factors can influence the impact of one or more SNP variants, increasing or decreasing the risk of illness. The MTHFR gene C677T polymorphism is an excellent example of a diet–SNP link. Because of this polymorphism, the MTHFR enzyme's activity is delayed. By folate, homocysteine is converted to methionine and then S-adenosyl-methionine, which is required for maintaining cytosine methylation in DNA and controlling gene expression (or folic acid). It is also possible to make a mutation that increases the amount of folate that can be used to make the DNA base thymidine while decreasing the incorporation of carcinogenic uracil (Hesketh, 2013). This change in methylation status may explain why homozygous C677T mutation carriers are more prone to developmental abnormalities while being protected from various cancers in low-folate environments. This change (for example, where there is a low intake of folate-rich vegetables such as spinach and asparagus or a lack of supplemental folate).

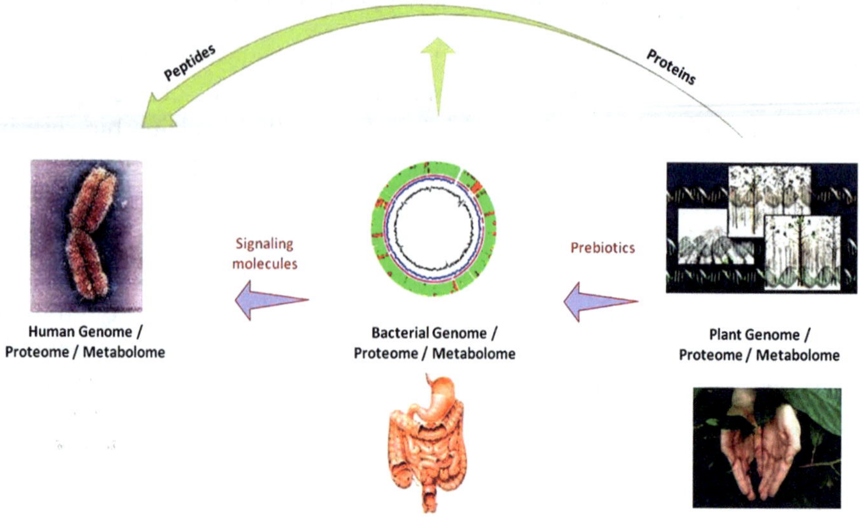

Figure 8.2: Digestive system.

Source: https://www.researchgate.net/figure/Extended-nutrigenomics-for-nutrition-and-health-the-plant-genome-for-example-affects-the_fig1_221804013

Folate (MTHFR's substrate) and riboflavin (MTHFR's cofactor) levels can influence MTHFR gene activity, which is significant. To put it another way, "depending on the fortification and supplementation strategies," the risks associated with MTHFR activity "may be significantly altered," according to Michael Fenech, a researcher at the CSIRO Genome Health and Nutrigenomics Laboratory. This technique may allow more neonates to be born with the MTHFR C677T polymorphism in countries where mothers are required to supplement with high-dose folic acid to avoid neural tube defects in their children. The process of converting folate into a usable form is more difficult for these children. These people may struggle to thrive if their diet is low in folate and riboflavin (Bionaz et al., 2015). High-throughput genomics (genomics, transcriptomics, proteomics, and metabolomics) would not have taken off if high-throughput genomic technologies had not recently become available.

"These technologies enable us to analyze multiple molecules of various types at the same time," says Jim Kaput, director of the FDA's newly established National Center for Toxicological Research (NCTR) Division of Personalized Nutrition and Medicine. "The discipline of genomics may contain SNPs and copy number variations in DNA, as well as a large number of unique RNA molecules." This is critical information because the majority of chronic illnesses are caused by complex interactions between multiple genes rather than single-gene changes. "These approaches can currently identify up to 500,000 SNPs per person. Nucleic acids, on the other hand, can be studied using either sequencing or hybridization methods, depending on the chemistry of the nucleic acid or protein under investigation. According to Kaput, the combination of several -omic technologies will result in a clearer picture of each individual's unique chemistry.

Nutritional biochemists are hard at work compiling data on the numerous nutrients and bioactive substances found in food that may be linked to molecular patterns discovered by various -omic technologies. Given the significance of studying genetic and nutritional trends in many populations and ethnic groups, the intersection of genomics and nutrition necessitates advanced analytic methodologies and, in Kaput's opinion, global free sharing of scientific research findings. Every day, human cells are bombarded with hundreds of DNA mutations, and if these errors are not corrected quickly, our entire genetic code will be erased. Diet and way of life are important factors in regulating this equation. According to a 2002 study published in Carcinogenesis, oxidative stressors such as smoking, strenuous exercise, and a high-fat diet aggravate DNA damage. The excretion of 8-ox-7,8-dihydro-2'-deoxyguanosine in the urine is reduced in people who eat a diet high in cruciferous vegetables and low in fat (8-oxide). Previously, vitamin C consumption was linked to the amount of 8-oxide in human sperm DNA, whereas fish oil and calcium consumption were linked to a lower incidence of oxidative DNA damage in colonic epithelial cells (Castro et al., 2006).

DNA and histone modifications, as well as genetic variations, are all important in maintaining genome integrity. Solo way believes that the loss of normal epigenetic states may cause "chromosomal rearrangements and increased failure of mismatch repair." "Because nutrients like folate have significant epigenetic influences, MTHFR alleles may regulate nutrient-related phenotypes through epigenetic mechanisms." Epigenetic changes in response to dietary factors frequently occur before genomic changes, though these genomic changes result in the generation of novel epigenetic patterns within the organism. It has recently been demonstrated that consuming antioxidant foods and phytochemicals can aid in DNA repair and protect against oxidative DNA damage, potentially compensating for genetic repair deficiencies. In those with genetic polymorphisms that reduce the activity of antioxidant enzyme systems such as manganese superoxide dismutase and glutathione peroxidase, a higher intake of antioxidants may be required to prevent DNA damage and cancer (Gonzalez et al., 2015). A single vitamin, such as folate, may be intriguing to research, but nutrition experts say that a wide range of dietary imbalances (both excess and deficiency) can have an impact on the genome. Fenech and colleagues found that the frequency of micronuclei in lymphocytes, a classic indicator of genome damage, increased in a study of 190 healthy men and women with an average age of 48 years. In May of that year, it was published in Carcinogenesis.

Previously, these dietary components were linked to lower cancer rates, osteoporosis, and hip fractures. They also tested the effect on folate intake when combined with calcium or riboflavin. Higher calcium consumption boosted the genome-protective effects of a high-folate diet, while higher riboflavin consumption exacerbated the harm caused by a low-folate diet. People who consume more red meat (high in riboflavin), drink more alcohol (which depletes folate) and consume fewer vegetables have a higher risk of cancer (a rich source of folate). Nutrition-modulated DNA repair techniques have piqued the interest of oncologists. According to gastroenterologist Graeme Young, director of Australia's Flinders Centre for Innovation in Cancer in Adelaide, dietary choices can help stabilize the genome after genetic aberrations appear. According to him, the traditional diet–genome method has linked dietary habits and the germline genotype to protection. "We're looking at the relationship between precancerous cells' aberrant DNA and nutrition." Young and his colleagues are currently researching DNA repair mechanisms.

8.2 BIOMARKERS OF NUTRITION AND CHRONIC DISEASE

According to Ben van Ommen, head of the European Nutrigenomics Organization, and his colleagues, inflammation, metabolism, oxidative stress, and psychological stress are all interwoven. One or more of these factors causes disease. Nutrigenomics is a significant endeavor to increase our understanding of the relationship between nutrition and genetics, at least in the first three categories. Nutritional genomics has the potential to play a significant role in the prevention of a wide variety of widespread modern illnesses, including obesity and diabetes, cancer and inflammatory disorders, and vitamin deficiencies. Medical anthropologists and others have named diabetes, obesity, and cardiovascular disease "civilization disorders." Adoption of a high-sugar, high-fat "Western diet" in indigenous cultures is frequently accompanied by a rise in obesity and diabetes, both of which are directly related to the new diet (Guéant et al., 2014). In-depth investigations of Arizona's Pima Indians and Hawaiian indigenous have found comparable conclusions. Diabetes, obesity, and cancer rates have increased due to the decline of the traditional plant-based, high-fiber diet.

Figure 8.3: Blood sugar level detector.

Source: https://www.diabetes.co.uk/blood-glucose/blood-glucose-testing.html

Diet has altered human metabolic capability throughout history, leading to the evolution of current disorders. According to evolutionary theory, diet, like other environmental elements, exerts selection forces on a population. If some genotypes in a population are linked to increased nutritional demands, selection will be performed against these genotypes if the needs are not satisfied. As long as these conditions are satisfied, the gene that transmits the high nutritional need will continue to exist in the population. This category may include genes linked to obesity and diabetes.

According to Soloway, when some gene variants give a selective advantage, high levels of an essential nutrient may lead to an increase in the frequency of specific alleles in a population. He believes that a population's genotype may alter as a result of selection acting on food availability. Obesity and diabetes are caused by a variety of factors, including an unbalanced diet and genes that originated during a time when food was scarce. The same genes are hypothesized to be responsible for a wide spectrum of harmful and maladaptive hormonal and metabolic features in today's culture (Carlberg, 2019). The chance of acquiring these illnesses is predicted to be impacted by genetic variations in sensitivity to the Western diet's effect on insulin resistance across ancestral cultures.

Lynn Ferguson, a nutrition specialist from New Zealand, says that "gene variants encoding taste receptors, or a range of peripheral signaling peptides like insulin, ghrelin, cholecystokinin, and corresponding receptors, have a considerable impact on food intake control." "The effect of this gene will have a significant impact on total food intake and the satisfaction value of specific meals." Researchers in the field of molecular diagnostics and therapy discovered a relationship between five widely detected single nucleotide polymorphisms (SNPs) and an increased risk of obesity and resistance to weight loss. These SNPs give a solid starting point for future nutrigenomic studies on people at risk of obesity, she says. These findings provide a strong case against a one-size-fits-all approach to combating the obesity pandemic. Obesity has been related to diabetes, heart disease, and a variety of malignancies, therefore treating the illness's nutrigenomic elements is crucial. According to research from the University of Navarra in Pamplona, Spain, study participants who had the Glu27 variant and consumed more than 49 percent of their total calorie intake were three times more likely to become fat (Chen & Zhao, 2013). As a consequence, when identical carbohydrate-to-calorie ratios were used, a variant form of the same gene was not associated with an increased risk of obesity. As a result, although some women gain weight on high-carb diets, others do not.

Regardless of obesity prevalence, abdominal fat remains a strong predictor of insulin resistance, type 2 diabetes, dyslipidemia, and cardiovascular disease. Polyunsaturated fat consumption was connected to reduced levels of belly fat in women who did not have a significant familial tendency to obesity, according to study participants at St. Vincent's Hospital in Sydney, Australia. Moderate alcohol use (1–1.5 drinks per day) was associated with a 20% decrease in abdominal obesity in women who were genetically prone to it.

Finally, nutrigenomics research advocates for an integrated strategy that combines genetics, public health, food science, and culinary arts to safeguard the human population from communicable illnesses. You'll get amazing outcomes if you add a little fat or butter to your diet. Though the job of preparing delicious and nutritious meals without the use of significant amounts of oil, butter, or other harmful fats is one that the entire community must face, it is required. The present trend of lifestyle disease involves the creation of a tailored nutrition meal plan based on an individual's genetic makeup by a nutritionist.

This chapter examines nutritional nutrigenomics in light of current advanced research to better understand its current and future potential and how to protect the human population from non-communicable illnesses. Diet has altered human metabolic capability throughout history, leading to the evolution of current disorders. According to evolutionary theory, diet, like other environmental elements, exerts selection forces on a population (Corella & Ordovas, 2009). If some genotypes in a population are linked to increased nutritional demands, selection will be performed against these genotypes if the needs are not satisfied. As long as these conditions are satisfied, the gene that transmits the high nutritional need will continue to exist in the population. This category may include genes linked to obesity and diabetes.

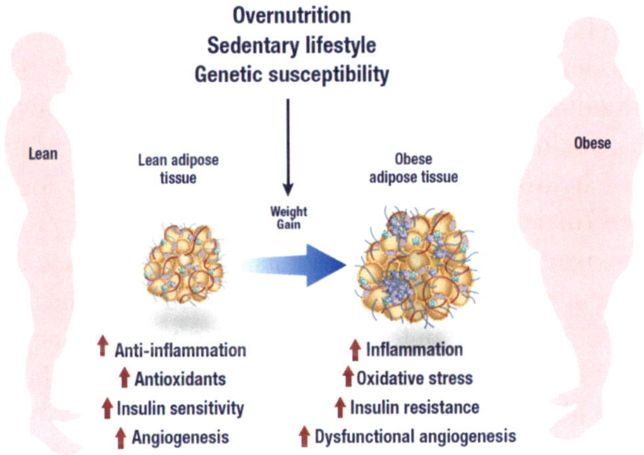

Figure 8.4: Effects of obesity.

Source: Dludla PV, Nkambule BB, Jack B, Mkandla Z, Mutize T, Silvestri S, Orlando P, Tiano L, Louw J, Mazibuko-Mbeje SE. Inflammation and Oxidative Stress in an Obese State and the Protective Effects of Gallic Acid. Nutrients. 2019; 11(1):23. https://doi.org/10.3390/nu11010023

Medical anthropologists and others have named diabetes, obesity, and cardiovascular disease "civilization disorders." Adoption of a high-sugar, high-fat "Western diet" in indigenous cultures is frequently accompanied by a rise in obesity and diabetes, both of which are directly related to the new diet. In-depth investigations of Arizona's Pima Indians and Hawaiian indigenous have found comparable conclusions. Diabetes, obesity, and cancer rates have increased due to the decline of the traditional plant-based, high-fiber diet.

8.3 NUTRITION DEFICIT AND BIOMARKERS

Nutrigenomics research focuses on the link between a person's diet and their genes, as well as approaches to protect humanity from noncommunicable diseases (NCDs) such as cardiovascular disease, obesity, diabetes, respiratory difficulties, metabolic syndrome, and malignancies worldwide. NCDs, also known as "urban junk food eating lifestyle disorders," are caused by long-term exposure to specific dietary components in urban regions. The use of biomarkers can help to identify these dietary deviations. Atypical lipid profiles, high blood pressure, or impaired insulin sensitivity may suggest the existence of noncommunicable diseases (NCDs) such as cardiovascular disease or metabolic syndrome. A person's genotype can impact the range of chronic illnesses that these biomarkers can detect (Fenech et al., 2011). These biomarkers are frequently single proteins, biomarkers, or biological processes that can detect changes in the body's proteome and metabolic systems. Individual DNA base damage may be detected using a variety of approaches, including oxidative stress-induced DNA adducts, such as the addition of hydroxyl radical to Guanine; (ii) abasic locations in the DNA sequence (measurable using the aldehyde-reactive probe); and (iii) DNA strand breakage (usually measured as deletions or base damage in the circular mitochondrial DNA sequence). Damaged DNA has recently been discovered to be a biomarker for a variety of disorders, including cancer, heart disease, and diabetes. The micronucleus test, which is the most reliable signal, can now identify DNA damage in cells with cytokinesis inhibited. The nutrigenomics technique known as transcriptomics has been carefully evaluated and confirmed to be trustworthy by using a microarray test to look at the copies of all actively transcribed genes (Connolly & Dawson, 2015). Hundreds of genes' expression levels may be measured simultaneously in a single experiment using this technology. Disease-specific gene expression patterns have been discovered in peripheral blood cells. Gene-expression patterns in blood cells that are specific to breast cancer and leukemia have recently been discovered and employed as disease indicators.

8.4 NUTRIENT SENSORS AND DIETARY CUES

Food has a wide variety of nutrient concentrations, and nutrigenomics can aid in deciphering the complexity of the human genome in reaction to these concentrations. Ruden et al. and their colleagues discovered that each nutrient has many targeting sites with varying affinity and specificity. Drosophila is the finest model organism for obesity and related illnesses because it contains adipose-like tissues and a lipid transport system that is similar to our own. As a result of their research into cell-sensing mechanisms, Müller and Kersten classified nutrients and metabolites as signaling components (Deb et al., 2014). Nutrients contain structural structures that include instructions on how to activate certain signaling pathways so that nutrients may reach their target spot. Changes in structure (such as saturated vs. unsaturated fatty acids or cholesterol vs. phytosterol) can have a significant impact on which sensor pathways are activated (e.g., saturated vs unsaturated fatty acids).

To understand the molecular pathways and upstream/downstream regulatory mechanisms mediated by diverse nutrients, scientists must labor long and hard. Nutrigenomics research enables the identification of molecular pathways through the genome-wide characterization of dietary target genes. Using this type of data, researchers may gain a greater understanding of the role nutrients play in health and sickness, as well as how they interact with food. This study will eventually result in evidence-based dietary and lifestyle recommendations, as well as nutritional therapies for today's population.

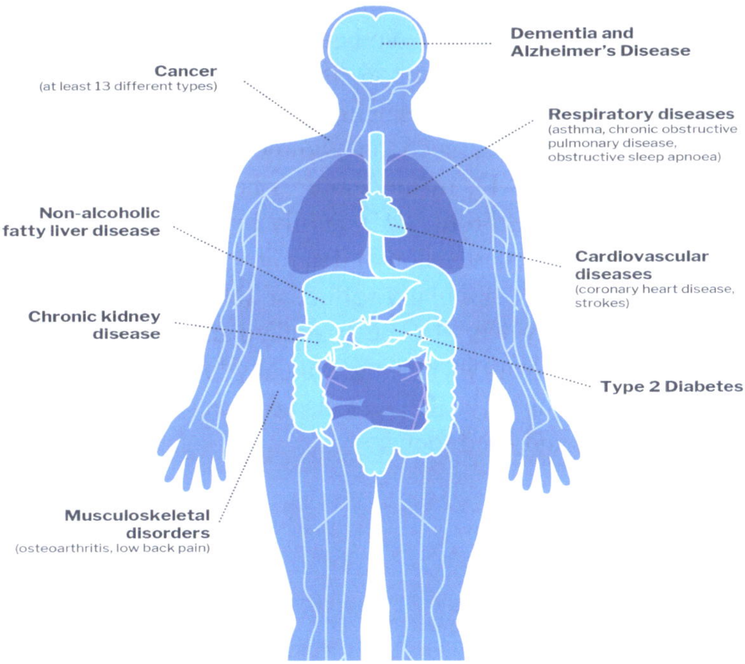

Figure 8.5: Obesity health impacts.

Source: https://www.obesityevidencehub.org.au/collections/impacts/health-impacts-of-obesity

8.5: THE RELATIONSHIP BETWEEN GENETICS, DIET, AND DISEASE

Nowadays, genomic diversity is most commonly defined by the use of single nucleotide polymorphisms (SNPs) or single nucleotide variations (SNVs) in genetic testing. Noncommunicable diseases (NCDs) such as cardiovascular disease, obesity, and diabetes can be avoided by following an adequate nutrition plan that considers a person's genotype. In Mexicans who consume a lot of carbohydrates and lipids, the sweet taste receptor (TAS1R2) and the cluster of differentiation 36 (CD36) have been related to dyslipidemia (Ferguson et al., 2010). People who take less folate, vitamin B6, and vitamin B12 and have frequent polymorphisms in homocysteine metabolism genes such as MTHFR and MTR (methionine synthase and methylene tetrahydrofolate reductase) have an increased risk of breast cancer.

The quantity of vitamin D in the population varies greatly, and it has the power to affect numerous metabolic processes in the body. SNPs in the vitamin D receptor (VDR) gene have been identified to change vitamin D availability and contribute to osteoporosis development in postmenopausal women who eat low quantities of calcium. SNPs were also found in genes that produce lipid proteins, such as apolipoprotein C3 (APOC3) and apolipoprotein A1 (APOA1). In moderate to heavy coffee users, genetic polymorphisms in the cytochrome P450 family 1 subfamily were also linked to an increased risk of hypertension and cardiovascular disease. The gene CYP1A2 (member 2): The cumulative effect of SNPs on dietary interactions and illness susceptibility has also been studied using GRS (genetic risk scores). Macronutrients with increasing adiposity levels can impact obesity GRS. In other studies, obesity GRS was also connected to the use of sugar-sweetened beverages and fried foods.

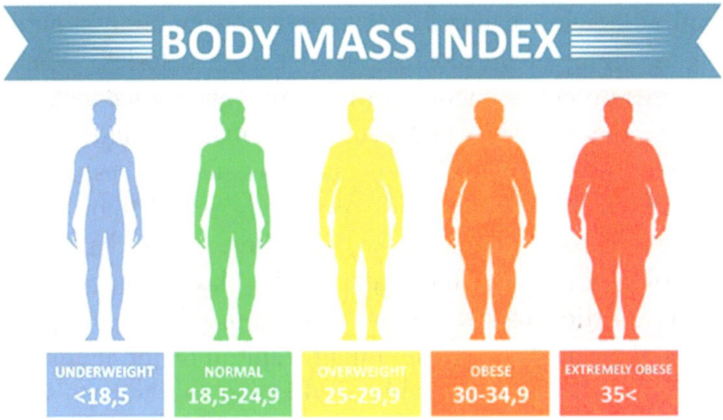

Figure 8.6: Body mass index.

Source: https://www.cdc.gov/healthyweight/assessing/bmi/adult_bmi/index.html

8.6 OBESITY AND NUTRITIONAL GENOMICS

Insulin resistance and hyperinsulinemia, hypertension, impaired glucose tolerance, noninsulin-dependent diabetic Mellitus, cardiovascular disease (CVD), type 2 diabetes, and a variety of cancers are all inflammatory comorbidities associated with obesity. To prevent obesity, nutrigenetic and nutrigenomics should be used in food and nutrition research, as well as lifestyle and nutrition research. Obesity does not impact everyone in an obesogenic environment; rather, genetic variability and the interaction of genetic and environmental factors decide whether or not a person or group is affected. The obesogenic environment is influenced by diet, age, gender, ethnicity, sleep duration, physical activity, sedentary behavior, stress, smoking, alcohol consumption, medication use, and depression. Based on the gene-nutrient-disease relationship, it is clear that environmental variables are the primary cause of obesity (Ferguson, 2016). This type of metabolic illness and its repercussions can be avoided by maintaining good eating habits and receiving at least 30 minutes of daily exercise. We may utilize nutrigenomics to investigate the numerous ways in which the genome and the regulatory systems it rules interact to generate obesity, a disorder that appears differently in various people.

Hill and colleagues claimed that energy consumption should exceed production, resulting in the deposition of excess calories. When energy intake surpasses output, a condition of positive energy balance is created, which increases the absorption and metabolism of dietary nutrients as well as the creation of energy through thermogenesis, resulting in a state of positive energy balance. To offset the effects of triacylglycerol lipolysis and fatty acid mobilization from adipose tissue, a positive energy balance must be achieved. Furthermore, environmental variables play an important influence in obesity. His assertion, made a century ago, was that the growing of embryos and fetuses in their mother's womb might result in major alterations in the progeny's genetic composition without affecting the embryo's or fetus's physical traits. One of the possibilities proposed in conjunction with these discoveries is epigenetic alterations connected to obesity.

Figure 8.7: Obesity health problem.

Source: https://www.lybrate.com/topic/what-are-the-health-risks-of-overweight-and-obesity/9638e6708921e81d26c617a324103ee6

The prenatal and early postnatal periods have a significant impact on obesity. In terms of epigenetics, the fetus's early nutritional environment has the greatest influence on the chance of obesity later in life. Epigenetics may impact gene expression without changing the DNA sequence; it is the key regulator and decisive factor in determining when and where specific genes are expressed. As established by Bird, DNA replication and cell division pass on a permanent epigenetic imprint left by methylation at the 5'position of cytosine in DNA enclosed inside a CpG (cytosine and guanine nucleotides linked by phosphate) dinucleotide (Fenech, 2008). This de novo methylation is maintained throughout mitosis by Dnmt1's gene-specific methylation of hemimethylated DNA, which is mediated by DNA methyltransferases 3a and 3b.

Furthermore, various studies conducted by several professionals explain a wide range of symptoms related to obesity. Nutrigenomic specialists can sometimes establish a newborn child's low-fat mass. Those born with a low birth weight are more likely to become obese later in life if they have early catch-up growth, which is defined by a greater accumulation of fat compared to lean body mass. If they are fed formula milk as a newborn, those born with a low birth weight are more likely to develop cardiovascular disease in adulthood. Many studies demonstrate that neonates who are fed formula milk instead of breast milk have a higher risk of obesity; nevertheless, certain persons do not meet this criterion. Cardiovascular illness primarily affects the heart and blood vessels (CVD). Atherosclerosis, capillary disease, and venous disease all contribute to cardiovascular disease.

Acute coronary syndrome (CAD) is a kind of heart disease that encompasses atherosclerotic, coronary, and ischemic heart disease, in which plaques grow on the inner artery walls and cause heart attacks. Although you cannot change your age, gender, or genetic composition, you may influence your risk of getting cardiovascular disease and how rapidly it advances. Hypertension, hyperlipidemia, obesity, diabetes, atherosclerosis, thrombosis, and smoking are all associated with an elevated risk of cardiovascular disease (CVD), and these risk factors are more prevalent than ever due to our fast-paced, high-stress lifestyles. After evaluating the complexities of the aforementioned CVD etiology, Juma et al. provide dietary advice for CVD prevention based on people's genetic constitutions (Dall'Asta et al., 2022). Dietary habits are thought to be environmental factors that are directly connected to the development of chronic diseases like cardiovascular disease (CVD). Researchers discovered that the content of a person's food is a significant risk factor for cardiovascular disease.

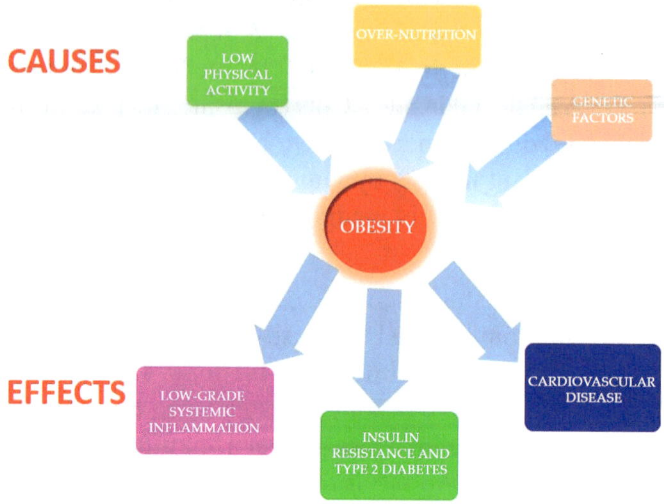

Figure 8.8: Bad habits that cause obesity.

Source: Gomez-Llorente MA, Romero R, Chueca N, Martinez-Cañavate A, Gomez-Llorente C. Obesity and Asthma: A Missing Link. International Journal of Molecular Sciences. 2017; 18(7):1490. https://doi.org/10.3390/ijms18071490

Obesity is a disease that affects the human race, putting people at risk for a variety of health issues such as heart disease, diabetes, and the list goes on. Certain "favorable" or "unfavorable" situations can contribute to the development of cardiovascular disease, which can be influenced by a variety of polymorphism genes. Lusis also highlights the role of atherosclerosis in the development of cardiovascular disease, arguing that it is the most significant factor and may be viewed as a complex mix of lipid transport and metabolic issues along with chronic inflammation. Despite a prolonged rise in total cholesterol, LDL cholesterol, and triglycerides in blood plasma, HDL (cholesterol) levels were discovered to have a protective impact against atherosclerotic illnesses.

The genes that generate apolipoproteins may be regulated by a range of signaling factors, such as hormones and enzymes, but this reveals that particular groups are more vulnerable to cardiovascular disease than others. Individuals with the E4 allele of the apolipoprotein E gene exhibited higher levels of low-density lipoprotein cholesterol after eating the same amount of fat as those with the E1, E2, and E3 alleles (bad cholesterol). An APOA1 gene promoter change from AG to A has been associated with greater

HDL-cholesterol levels, however, the results have been inconsistent among studies. In contrast, Ordovas observed a relationship between the A allele and decreased HDL levels in the blood. However, PUFA intake restored the genetic influence in women.). When alcohol and cigarette smoking were included, the effect of this type of fat on males was amplified (Dang et al., 2014). A variety of gene variations encoding lipid transport proteins, receptors, enzymes, and inflammation-related genes have been associated with a variety of blood lipid diseases.

The correct macronutrient ratio, total fat content, and specific fatty acid content are critical in the prevention and treatment of cardiovascular disease. The composition of monounsaturated fatty acids in a person's diet varies substantially depending on whether it comes from olive oil, meat, or any other meal including an animal source. Furthermore, as Jacobsen et al. and Russo point out, there is much disagreement on whether polyunsaturated fatty acids (PUFAs) are optimal for heart health and preventive or therapy (omega-6 and omega-3 series). Both Shai et al. and Sacks et al. used the same platform and case study to underline the importance of a high carbohydrate, low-fat diet for regulating body weight and cardiovascular risk factors. According to their 1965 study, an individual's "intrinsic features" can impact the effect of diet on plasma cholesterol concentrations (Fallaize et al., 2013). To avoid CVD, nutritional counseling focuses on weight loss and lipid profile stabilization through diet, exercise, and medication.

8.7 NUTRIGENOMICS AND TYPE 2 DIABETES

Diabetes mellitus (DM), a collection of metabolic illnesses defined by insulin malfunction, insulin secretion failure, or both, is characterized by hyperglycemia. Diabetes mellitus (DM), a metabolic condition that affects numerous organs, including the heart, blood vessels, and kidneys, is increasingly being viewed as a worldwide burden by experts. According to the International Diabetes Federation, diabetes is anticipated to affect 592 million people in the next 25 years, up from 8.3 percent of adults (382,000,000) today. T1DM and T2DM are the etiopathogenetic classifications for type 1 diabetes and type 2 diabetes, respectively. T2DM, he said, will be a serious health issue over the world, putting a strain on health-care expenditures. Obesity more than tenfold increases a person's chance of having diabetes. People in developing nations are increasingly eating a diet heavy in processed carbs and red meat, which contributes to obesity. Insulin, a hormone that controls glucose and fat metabolism, is produced by pancreatic cells (Elsamanoudy et al., 2016). Insulin resistance has been related to both obesity and type 2 diabetes. Regular ingestion of sugar and saturated fat causes glucolipotoxicity, which limits insulin production by beta cells, resulting in hyperglycemia and high cholesterol.

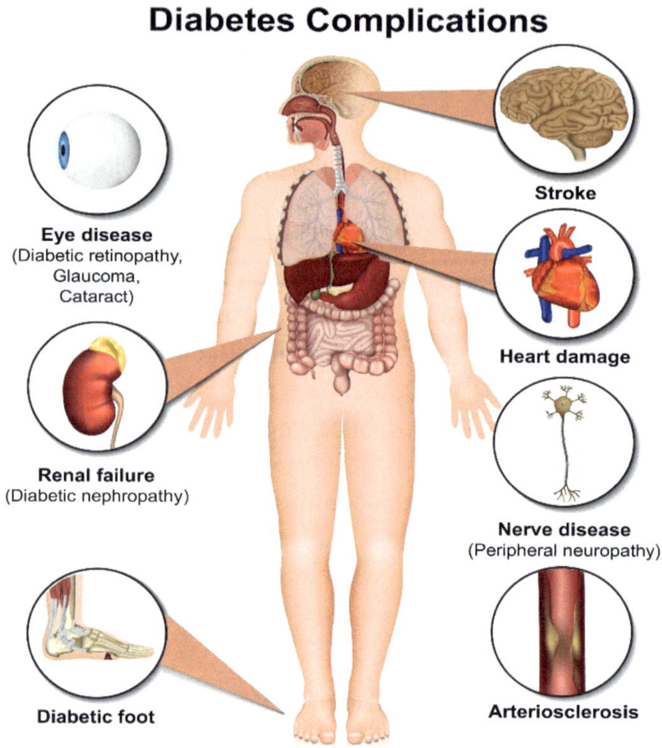

Figure 8.9: Diabetes complications.

Source: https://www.medicdrive.org/diabetes/complications-of-diabetes/

Anthocyanins, flavones, flavanols, flavanones, and isoflavones are examples of these. Polyphenolic compounds, such as flavones, luteolin, and quercetin, have a significant impact on intracellular signaling pathways in the human body (insulin secretion, insulin signaling, and glucose uptake; improving mitochondrial status; suppressing inflammatory cytokine production and reactive oxygen species (ROS)/reactive nitrogen). In addition to flavonoids, phenolic acids and tannins have been demonstrated to block the carbohydrate-breaking enzymes -glucosidase and -amylase. People who ate apples or drank tea were shown to have a lower risk of developing type 2 diabetes. According to the National Association of Health Professionals, the United States has around two million health professionals (Dimitrov et al., 2016). Follow-up research has found a relationship between anthocyanin-rich foods including blueberries, apples, and pears and a decreased risk of type 2 diabetes.

The most prevalent phenolic acids (found in cereals and converted to hemicelluloses) are caffeic acid, chlorogenic acid (found in coffee), and ferulic acid (found in cereals and converted to hemicelluloses). According to various epidemiological research reports, persons who consume up to six cups of coffee per day have a much-reduced chance of getting T2DM than those who do not consume any coffee at all.

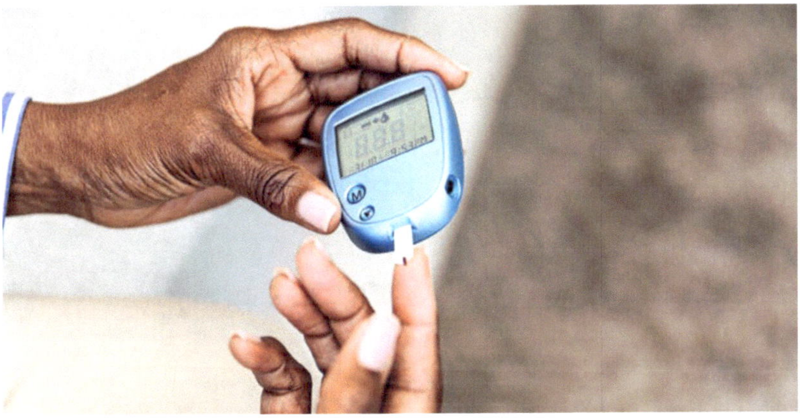

Figure 8.10: Blood sugar tester.

Source: https://www.insider.com/guides/health/how-to-check-blood-sugar

Resveratrol (trans-3, 5, 4′-trihydroxystilbene), a phytoalexin, has been proven in animal models of diabetes and human patients to help slow the progression of diabetic problems in numerous organs and tissues. Furthermore, it helps to maintain pancreatic cells and glucose balance. It aids in insulin synthesis in the treatment of metabolic diseases. People who are vitamin D deficient are more prone to develop type 2 diabetes (Dimitrov, 2011). When combined with high levels of parathyroid hormone (PTH), hypo vitamin D levels are an independent predictor of -cell dysfunction, insulin receptor dysfunction, and glycemia. Vitamin D treatment improves glycemia and insulin secretion in T2DM patients with proven hypovitaminosis not only by directly impacting pancreatic-cell function but also by modifying plasma calcium levels, which influence insulin synthesis and secretion.

8.8 CANCER AND NUTRITIONAL GENETICS

Cancer development is influenced by genetic, protein, and metabolite abnormalities. Food manipulation has resulted in a plethora of information on biological pathways that cause carcinogenesis, which has aided the genomic era in understanding this disease. Inherited gene mutations can raise a person's risk of cancer. Genetic and dietary variables may impact cancer risk (Dimitrov, 2011). Endogenous responses are known to cause cancer include aflatoxin (a carcinogen linked to liver cancer) and extremely high doses of ionizing radiation (many forms of cancer).

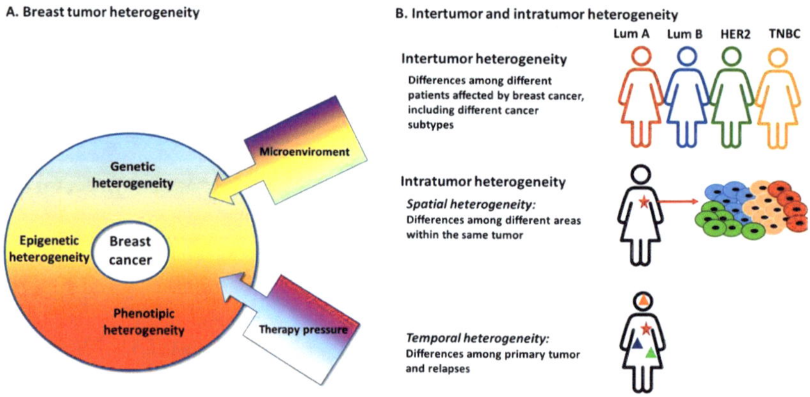

Figure 8.11: Breast cancer illustration.

Source: Fumagalli C, Barberis M. Breast Cancer Heterogeneity. Diagnostics. 2021; 11(9):1555. https://doi.org/10.3390/diagnostics11091555

If homeostasis is to be maintained, nutrition, metabolism, and gene expression must all communicate with one another. Human diet and human DNA collaborate to produce the same effect. This may have an impact on a person's health and susceptibility to illness. Nutrients regulate transcription factors, which impact gene expression (up or down), which in turn influences metabolic response. All of the diet's protective, carcinogenic, and mutagenic substances are digested by biotransformation enzymes. Gene variations that change protein expression or enzyme activity can increase the risk of cancer. More than 25,000 such molecules are believed to be found in human diets. More than 500 distinct bioactive food elements have been discovered as a result of research into the involvement of bioactive food components in cancer genesis.

Food is regarded as a protective mechanism for carcinogens as a source of carcinogens (intrinsic or cooking-generated) (vitamins, antioxidants, enzyme activating compounds, etc). Carcinogen metabolism polymorphisms may modify the chance that carcinogens may interact with target cells, hence interfering with the onset of cancer (Wiwanitkit, 2012).

Polymorphisms in genes encoding hormone-related components are especially common in hormone-dependent malignancies such as breast, prostate, ovarian, and endometrial cancer. According to research published in the New England Journal of Medicine, genes that code for estrogen, progesterone, and androgen receptors have been associated with a lower risk of cancer in women. Hormone regulation can be influenced by a variety of dietary variables. Obesity has a substantial impact on hormonal balance. Certain investigations have indicated that the metabolism of phytoestrogens and other dietary components is similar to that of sex hormones.

Figure 8.12: Healthy living.

Source: https://vitagene.com/blog/what-is-nutrigenetics-and-nutrigenomics/

A person's diet, for example, can have an impact on their cancer risk. High red meat consumption increases the risk of colon cancer. N-acetyl transferase (NAT) is a phase II metabolism enzyme that exists in two forms: NAT1 and NAT2. Some NAT1 and NAT2 polymorphisms can have slow, moderate, or quick acetylation. Heterocyclic aromatic amines, like those found in NAT, are typically present in the air when heated, as in the case of cooked red meat. Certain amino acids may react with creatinine during high-temperature cooking of muscle to form heterocyclic aromatic amines

(HAA). Acetylation of HAA results in the formation of cancer-causing reactive compounds. This acetylation can only be performed by NAT2 rapid acetylators. Those who ate a lot of red meat and had the NAT fast acetylators genotype were more likely to develop colon cancer.

According to a study, salts and preservatives are hazardous for stomach cancer. Colon cancer risk is reduced in people who have the MTHFR gene variant C667T, which reduces enzyme activity. A diet deficient in folate, vitamin B12, vitamin B6, and methionine enhances the risk of cancer in people with the MTHFR CC or TT phenotype (Alagawany et al., 2022).

Reactive oxygen species (ROS) such as superoxide anions, hydrogen peroxide, and hydroxyl radicals have been shown to change DNA sequences. When DNA replication is disturbed, oncogenes and tumor suppressor genes can be changed. ROS can induce DNA strand breaks, which can result in genetic mutation or deletion.

The fiber in the diet may help lower the chance of getting colon cancer. Fish oil, which is abundant in omega-3 fatty acids has been demonstrated *in vitro* and *in vivo* to lessen the chance of developing colon cancer.

Several bioactive compounds contained in fruits and vegetables, such as those that limit metabolic activity and improve detoxification, can aid in cancer prevention. Indoles, flavonoids, phenols, isothiocyanates, and allyl sulfur compounds, as well as selenium, can all affect the body's detoxification enzymes.

Carotenoids, flavonoids, and allyl sulfur compounds, as well as conjugated linoleic acid, conjugated N-3 fatty acids, and calcium, are bioactive components that can influence carcinogen metabolism, cell signaling, cell cycles, apoptosis, hormonal balance, and angiogenesis. Lycopene, resveratrol, hesperidin, ascorbic acid, coffee acid, soluble fiber, polyunsaturated fatty acids, and resveratrol, which may be found in tomatoes, as well as resveratrol, which may be found in grapes and berries, are cancer inhibitors. Diets rich in minerals and vitamins that control enzymatic processes have been demonstrated to reduce cancer risk (Wittwer et al., 2011). Micronutrient deficiencies might result in abnormalities. Zinc and folate, for example, are involved in the DNA repair process. Other natural plant substances that have been demonstrated to impact the cell cycle include apigenin, curcumin, epigallocatechin-3-gallate, resveratrol, genistein, and allyl sulfur. This is consistent with prior research findings. Some of these alterations, such as changes in the phosphorylation of critical regulators of cell division, may be caused by the post-translational processing of newly generated proteins. Other dietary components can influence tumor activity by enhancing apoptosis and hastening cell death. The intrinsic and extrinsic apoptosis routes, both of which are mediated by the mitochondria, are well-known apoptosis processes

Figure 8.13: World cancer day.

Source: By Lala sneh - Own work, CC BY-SA 4.0, https://commons.wikimedia.org/w/index.php?curid=58227513

The American Cancer Society has shown that vitamin C-rich foods, such as fruits and vegetables, reduce the risk of cancer in multiple trials. Supplementing with vitamin C, on the other hand, appears to not influence cancer incidence. According to research, the potential of fruits and vegetables to prevent cancer stems from a mix of vitamins and phytochemicals, rather than simply vitamin C.

Specific micro-and macronutrient absorption, for example, may be linked to chronic disorders caused by SNPs. The authors of this study show how certain SNPs in certain ethnic groups can cause illness (Van Breda et al., 2015). Due to SNPs in genes coding for "sweetness" taste receptors, people of Mexican ancestry have an increased risk of dyslipidemia and elevated cholesterol when compared to people of other ethnicities.

Heavy coffee drinkers who have a mutation in the gene that codes for certain cytochrome p450 enzymes, which are essential to the citric acid cycle and move electrons across membranes to produce energy, are more likely to develop cardiovascular disease and high blood pressure. According to some studies, drinking three to five cups of coffee per day may reduce the risk of cardiovascular disease and hypertension, but the exact link between coffee and cardiovascular disease remains unknown. SNPs in the FTO gene, which codes for an enzyme that demethylases DNA (activates or deactivates genes), may result in weight loss or higher-than-normal insulin levels when following a high-protein or low-fat diet. Individual nutrition has the potential to produce far greater health benefits than dietary recommendations based on general population trends. By researching a person's DNA and removing or adding specific meals, we can improve outcomes.

Because of individual differences in absorption, distribution, metabolism, and disposal, dietary intake of a nutrient may not always result in the same blood or tissue concentrations. Individual differences in food response pathways are extremely complicated and remain a mystery to scientists. Diet-gene interactions influence inter-individual variability, which has aided in the identification of some of these genetic variations thus far. The interaction of genetic and dietary variables can increase the risk of illness in some people and communities. Diet-gene association data can be used to support gene-specific dietary intervention trials that are currently underway. The current findings must be replicated, and additional research using genotype-specific dietary intervention trials is required. Future nutrigenomics research could help us better understand biological function and individual dietary responses to a variety of foods (Su & Lu, 2012). A path to a personalized diet may be paved using this iterative approach to health research. The use of nutrigenomic techniques to diagnose some monogenic disorders, such as PKU, is an efficient use of current health-care initiatives, such as newborn screening programs. Nutrigenomics has yet to gain widespread acceptance among medical professionals as a method of preventing and treating complex and debilitating chronic illnesses.

Although such treatment is unlikely to become available to the general public anytime soon, nutrigenomics theories and techniques are expected to allow for earlier and more precise treatments in the future Because current research examines how diet-gene interactions impact the phenotypes identified as predictive biomarkers of disease, these chronic illness indicators will almost certainly be included as outcome measures in future nutrigenomics research Because many previously unknown etiologic pathways in disease pathogenesis have been discovered, genome-wide techniques have the potential to discover new dietary targets for illness prevention or therapy The diet of an individual is an important environmental factor that influences the likelihood of developing the illness. A better understanding of these connections could aid in disease prevention. The extent to which nutrigenomics will be integrated into nutrition therapy and promotion remains to be seen. Nutritional genomics is a new field of study that has the potential to change how future dietary guidelines are developed and recommended for communities and individuals.

Lactose intolerance is caused by a genetic mutation, and PKU is caused by a dietary factor. The challenge for nutrigenomics is to identify gene-nutrient interactions that influence complex polygenic disorders like obesity, diabetes, cancer, and cardiovascular disease, which can take years or even decades to develop and have multiple underlying causes. This makes determining the role of specific food components and genetic variants in the onset of specific diseases difficult. Diet-gene interactions occur when a genetic factor influences the effect of a dietary component on a specific phenotypic or health outcome measure, such as high blood glucose or obesity. The influence of dietary change on these variables may be related to nutritional interactions between genes and phenotype or health outcomes (Ordovas 2008). The primary goal of nutrigenomics is to prevent the onset and progression of chronic disease. To achieve this goal, research projects are now expanding the corpus of knowledge linking diets to metabolic pathways that influence disease outcomes. Genetics is being used in nutritional epidemiology research to try to improve the consistency of the results. People and specific subpopulations may benefit from the new research, which may result in the development of personalized dietary advice and, as a result, a lower risk of chronic illness.

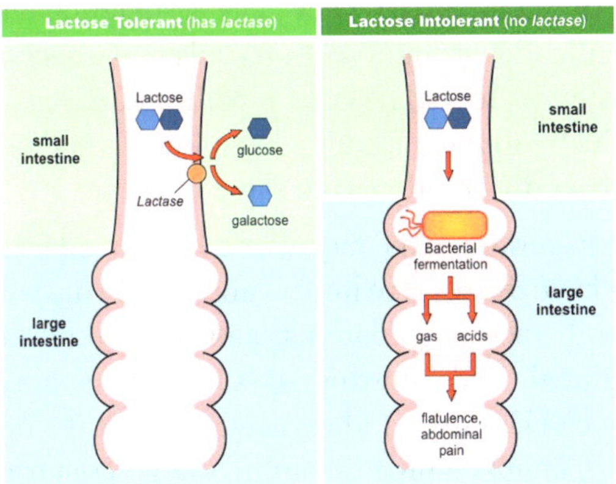

Figure 8.14: Lactose intolerance.

Source: https://ib.bioninja.com.au/standard-level/topic-2-molecular-biology/25-enzymes/lactose-intolerance.html

In nutrition studies, genetic variability data can be used to determine a food's link to a specific health outcome. This link between food and disease is established by examining whether functional changes in genes involved in dietary component metabolism alter the illness-diet relationship. Moderate alcohol use, for example, has been linked to a decreased risk of heart disease. The rationale for this link was that moderate drinkers had higher levels of high-density lipoprotein (HDL) cholesterol. Although moderate alcohol consumption is protective, this may be due to a lifestyle trait that distinguishes moderate drinkers from abstainers and heavy drinkers(Singh & Sharma, 2019).

Nutrition and genetics both play a role in the development of chronic diseases in the human population, including cancer, osteoporosis, diabetes, and cardiovascular disease. Nutrigenomics investigates the impact of diet on human health by incorporating high-throughput "omics" technologies such as transcriptomics, proteomics, and metabolomics. Because of individual differences in absorption, distribution, metabolism, and disposal, dietary intake of a nutrient may not always result in the same blood or tissue concentrations. Individual differences in food response pathways are extremely complicated and remain a mystery to scientists. Diet-gene interactions influence inter-individual variability, which has aided in the identification of some of these genetic variations thus far. The interaction of genetic and dietary variables can increase the risk of illness in some people and communities (Riscuta, 2016). Diet-gene association data can be used to support gene-specific dietary intervention trials that are currently underway. Individual vulnerability to sickness or health and dietary response is just 0.1 percent genetically diverse from one another. As if that weren't enough, different people react differently to the same meal.

Aside from that, certain diets may alter genes to aid health maintenance, whilst others may have the reverse impact and induce disease. Individualized nutrition and food recommendations can be produced based on an individual's nutritional needs, nutritional status, and genotype to maintain a good balance between health and illness and to support a healthy lifestyle. A tailored nutrition system is a pipe dream in a large country like ours, where individuals are still struggling to satisfy their fundamental necessities. Noncommunicable illnesses are no longer a threat to rural India, which has begun to embrace a more urban lifestyle. Rural India's culinary preferences have altered dramatically since the globalized age of urbanization.

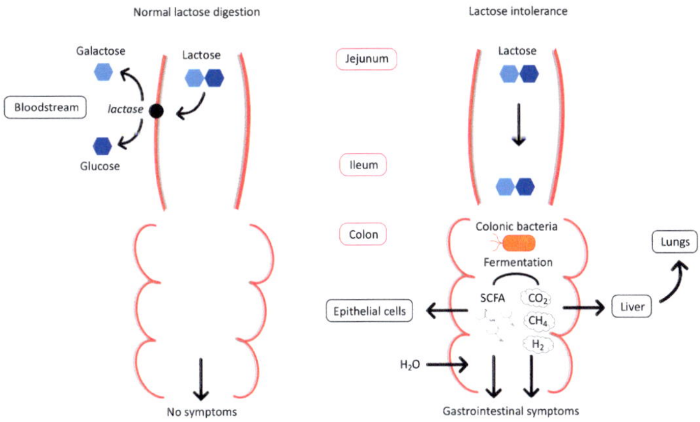

Figure 8.15: Lactose intolerance process.

Source: Porzi M, Burton-Pimentel KJ, Walther B, Vergères G. Development of Personalized Nutrition: Applications in Lactose Intolerance Diagnosis and Management. Nutrients. 2021; 13(5):1503. https://doi.org/10.3390/nu13051503

The most critical problem right now is determining how to address this public health concern. Even though nutrigenomics is the only technology that works, it is extremely expensive for the typical individual to afford. Dietary intervention for so many individuals demands a large workforce and cutting-edge scientific facilities, making it a difficult task (Riscuta & Dumitrescu, 2012). These ailments can be avoided in several ways by raising public knowledge about them. It will emphasize early identification of at-risk individuals and appropriate intervention in the form of weight loss, diet adjustments, and increased physical activity to avoid or at least postpone the formation of dietary disorders.

CHAPTER 9

REGULATION AND POLICY CONSIDERATIONS OF NUTRIGENOMICS

CONTENTS

9.1 INTRODUCTION

Nutrigenomics is an emerging discipline that examines the role of genetic variations in the interaction between nutrients and gene expression. This field has far-reaching implications for a number of industries, including agriculture, food processing and pharmaceuticals. Consequently, it also has regulatory implications for all of these fields as well. The FDA, USDA and EPA have yet to set any formal policy on nutrigenomics research or regulation because the technology is still so new and developing at such a rapid pace (Ray, 2014). However, there are some fundamental issues they should consider when setting their policies in this field going forward.

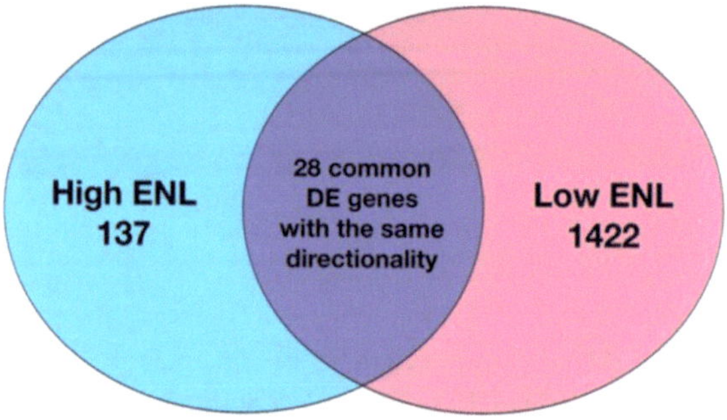

Figure 9.1: A Venn's diagram overview of Nutrigenomics for lignin in enterolactone (ENL) and enterodiol (END) phenotypes.

Source: Mullens DA, Ivanov I, Hullar MAJ, Randolph TW, Lampe JW, Chapkin RS. Personalized Nutrition Using Microbial Metabolite Phenotype to Stratify Participants and Non-Invasive Host Exfoliomics Reveal the Effects of Flaxseed Lignan Supplementation in a Placebo-Controlled Crossover Trial. Nutrients. 2022; 14(12):2377. https://doi.org/10.3390/nu14122377

The field of nutrigenomics is the study of how food and nutrition affect gene expression. It draws from both fields: genetics and nutrition. In fact, it is important to note that genetics does not change our genes in any way; rather, it influences how our genes are expressed. Nutrigenetic testing uses data obtained from blood samples or saliva tests to determine whether a patient's diet has an impact on their genetic predisposition for certain diseases or conditions (1). The number of available tests has been growing since 2016 when Dr. David Ludwig published his book "Always Hungry?" where he discussed how our diet can directly alter our genes (Qadeer & Ashhar, 2014).

Regulation and policy considerations on Nutrigenomics is a special issue of Regulatory Toxicology and Pharmacology, focusing on the key factors impacting research, development and approval of foods with health claims.

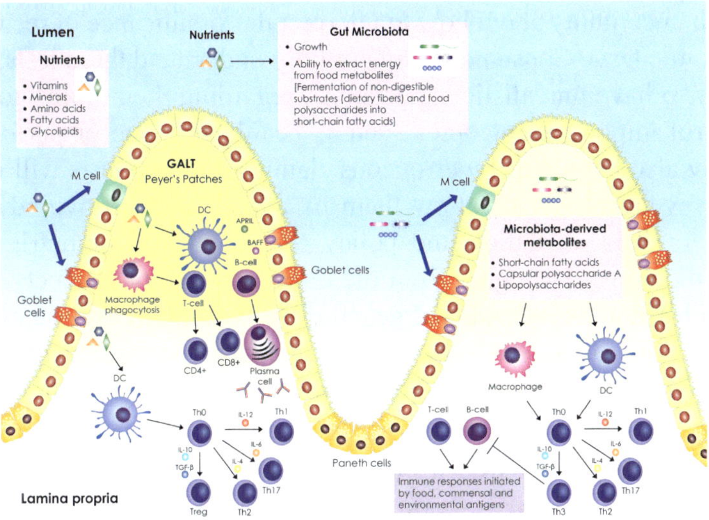

Figure 9.2: Interlink and functionality of Nutrigenomics.

Source: Tourkochristou E, Triantos C and Mouzaki A (2021) The Influence of Nutritional Factors on Immunological Outcomes. Front. Immunol. 12:665968. doi: 10.3389/fimmu.2021.665968

Regulation and policy considerations on Nutrigenomics are very important for public health. The lack of regulation has resulted in an explosion of food products making health claims about their ability to modify the DNA; however, there are many unresolved questions about the safety of these foods and the value of their claims. In addition, it is unclear whether companies can privately test their own product samples and make claims directly or must register with a regulatory authority before being able to make such claims. There is a need for an appropriate legal framework and clear policy for nutrigenomic testing, however. This will help protect against exploitation by private enterprises, particularly data leakages and unauthorized access. The regulation and policy regimens should be established by governmental organizations such as the Food and Drug Administration (FDA) which ensures the manufacturing, distribution, and marketing of foods, prescription drugs, over-the-counter medicines and other products (Sanders et al., 2015).

Regulation of the field remains a challenge for many reasons. The science is new and rapidly developing, with many unanswered questions and much uncertainty about how to interpret the significance of results from nutrigenomic tests. Consumers may not fully understand these tests, so they are likely to have unrealistic expectations regarding their ability to predict and control important outcomes such as health, pleasure or performance. They may also be misled by erroneous claims that these tests will improve health, prevent disease or allow them to achieve competitive advantages over others. The regulation and policy considerations for nutrigenomics are extremely broad, ranging from the responsibility for product claims to the consideration of individuals' genetic information. There are numerous regulatory issues that need to be considered when developing nutrigenomic products from regulatory issues on health claims, attributes, ingredients and manufacturing processes.

The possibilities for use of nutrigenomic techniques and personalized nutrition are likely to increase considerably in the coming decade. Regulatory, legal, economic and ethical issues must be considered as personalized nutrition becomes increasingly available through scientific advances, public health initiatives or commercial applications. The regulation of nutrigenomic therapies will require consideration of both the benefits and risks to patients. Although nutrigenomics already plays an essential role in clinical practice, we need to accommodate new technologies and services as they develop. As the science and practice of nutrigenomic research advances, healthcare professionals have a responsibility to educate their patients about the implications, benefits and risks associated with this emerging science (Pauletto et al., 2020). It is hoped that this will provide opportunities for individuals to make choices regarding their genetic tests and interpretation of test results based on their own priorities.

The US Food and Drug Administration recently updated their public information on the regulation of personalized nutrition products, including genetic tests sold directly to consumers. This policy update includes a summary of the regulatory history and current issues, discussions of the potential consumer benefits from improvements in accuracy, a report of FDA efforts to understand consumer perception regarding the benefits and risks of direct-to-consumer testing for disease risk and prevention, recommendations for FDA guidance and future rulemaking to ensure adequate consumer protection while allowing innovation in personalized nutrition. This shows that the science of nutrigenomics is considered an important scope in the field of health and medicine.

9.2 RESEARCH AND DEVELOPMENT ON NUTRIGENOMICS

The research and development on nutrigenomic aims to discover and improve the risk assessment of metabolic disorders, cardiovascular diseases and cancer by analyzing the interaction between a single food ingredient and several different genetic variants. Furthermore, Nutrigenomics researches the role of nutrition in regulating gene expression, protein activity and metabolism. Research projects on nutrigenomic are conducted to understand how a nutrient diet can alter the genome function, which in turn affects disease risk and health outcomes.

Nutrigenomics is the study of how nutrition affects gene expression. It is a relatively new field, but it has already been shown to have a significant impact on human health. In particular, it can be used to understand how gene-diet interactions affect disease and health. Research into nutrigenomics focuses on three main areas: 1) understanding how food interacts with our genes to influence health; 2) developing methods for detecting abnormal genes in order to predict disease risk; and 3) developing methods for correcting abnormal genes in order to prevent disease (Peña-Romero et al., 2018). One of the most important aspects of research done on nutrigenomics is the use of genomic information to understand gene-diet interactions in humans as an aspect of research done on nutrigenomics.

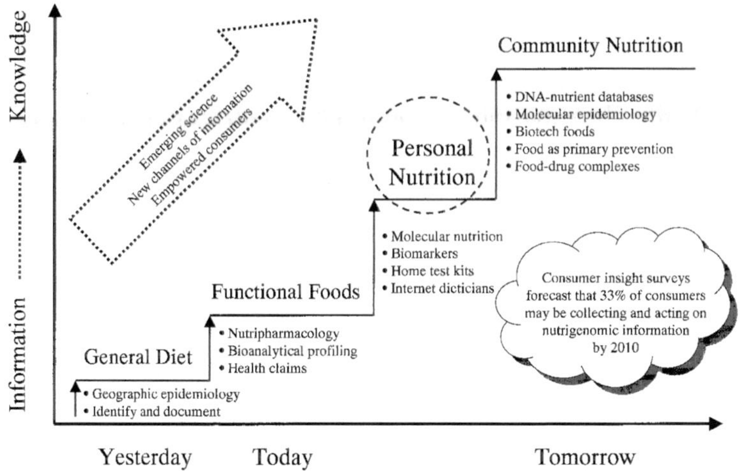

Figure 9.3: An overview of the existing and future recommendations on Nutrigenomics.

Source: https://www.semanticscholar.org/paper/Nutrigenomics%3A-the-Rubicon-of-molecular-nutrition.-Gillies/da2436513bacacf38f505e8f3ab2d2f82d5bf893

It's no secret that many factors influence our health, including genetics, lifestyle, and environment. But a lesser-known factor is the food we eat. Food interacts with our genes to influence us in many ways—for example, by affecting how well our genes function or by changing the expression (or lack thereof) of certain genes. This interaction can be very beneficial for our health if we choose foods that work with our bodies' genetic makeup to promote health. But if we choose foods that do not work with our bodies' genetic makeup, they may have negative effects on us and even cause disease (Afman & Müller, 2006).

The human genome is made up of a sequence of DNA, which is the genetic material that makes up our genes. Genes are the units of heredity that carry the information for making proteins, and proteins make up our bodies. The human genome consists of 23 pairs of chromosomes, with each chromosome carrying tens of thousands of genes. Each gene has a specific location on a specific chromosome, and there are about 20,000 genes in total. This means that every person's genome is unique, and contains roughly 3 billion base pairs (or nucleotides). The discovery of DNA structure by Watson and Crick in 1953 allowed scientists to start studying how genes work in more detail. They discovered that DNA consists of two twisted strands held together by hydrogen bonds between complementary bases—A bonds with T, and C bonds with G. These strands can be unwound when exposed to heat or electricity, allowing them to be copied into RNA molecules by enzymes called polymerases. These RNAs then direct protein production inside cells through the use of ribosomes—protein factories present in every cell that read information from messenger RNAs (mRNAs) transcribed from DNA sequences during transcriptional processes occurring at all times within cells throughout life. With this research, one can easily understand the human genome and the risks that it may encounter thus counteracting the negative impacts that may lead to serious health issues.

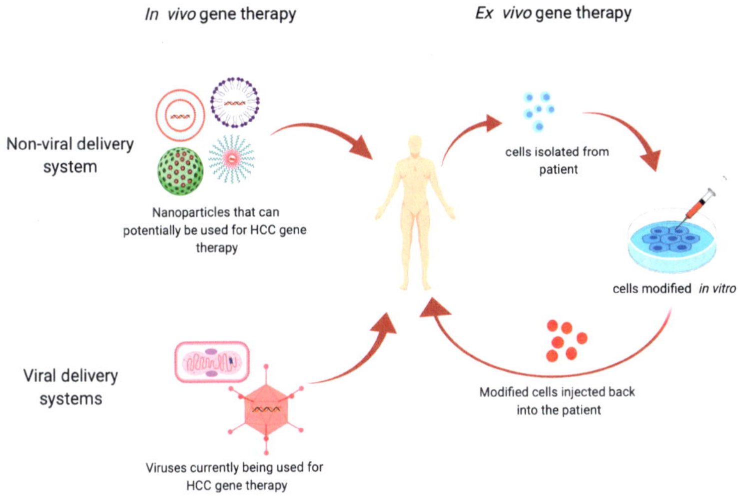

Figure 9.4: Genome editing; one of the ways of correcting abnormal genes.

Source: Reghupaty SC, Sarkar D. Current Status of Gene Therapy in Hepatocellular Carcinoma. Cancers. 2019; 11(9):1265. https://doi.org/10.3390/cancers11091265

One of the most promising ways to fight disease is by correcting abnormal genes in order to prevent disease. This has been demonstrated by the success of genome editing. In this process, we use a tool called CRISPR-Cas9 that essentially acts like a pair of scissors to cut out specific genes and replace them with healthy ones. The technology has been used successfully in mice, zebrafish, and human embryos. It's also been used on adults to treat diseases like Huntington's and sickle cell anemia (Banerjee et al., 2015).

There are two primary methods for correcting abnormal genes that lead to disease. The first is gene therapy, which involves using genetic engineering techniques to replace or add a gene that has been lost or damaged. The second is genome editing, which involves editing the DNA sequence in question (typically by introducing small changes) to correct an error that caused the disease in the first place.

Nutrigenomic product development is an emerging field of scientific research that investigates how food and nutritional products can be designed for optimal metabolic health. SNPedia and NutriGenome Central are two databases to help researchers discover correlations between genes, our environment and the foods we eat. The goal of nutrigenomics is to develop biomarkers that are associated with diet and then identify promising targets for therapeutic development. The focus is on new treatments, the identification of novel disease mechanisms, and the development of personalized therapies with minimal side effects.

A list of research and development aspects done on nutrigenomic which includes genome-wide association studies, genetic variants associated with nutritional status and diseases, study of the relation between nutrition and biomarkers, and use of genomic information to understand gene-diet interactions in humans. The purpose of genome-wide association studies (GWAS) is to identify the genetic variants that are consistently associated with a particular phenotype, disease or trait. By doing so, GWAS can be used to identify genes and pathways that play important roles in the development or prevention of diseases. GWAS is a powerful tool to identify genetic variants that are associated with complex traits and diseases. The ability to conduct large-scale GWAS has opened new opportunities for identifying common genetic variants associated with complex traits and diseases (Benítez et al., 2017). Nucleic acid sequencing has been used to study gene expression on a genome-wide scale. One approach, genome-wide association studies (GWAS), look at all genes simultaneously and compares the DNA sequences of individuals with a particular disease or condition to tell us what part of the genome is involved in these diseases. The first GWAS was conducted by the Wellcome Trust Case Control Consortium in 2007 and identified variants at more than ninety genomic regions associated with human diseases. The impact of this discovery was enormous and quickly led to a flood of follow-up investigations from groups around the world over the next decade.

A variety of genetic variants have been associated with nutritional status and diseases. These include single nucleotide polymorphisms (SNPs) near genes that encode proteins involved in energy metabolism, protein synthesis and degradation, homeostasis of polymers and other mitochondrial proteins, as well as inflammatory responses. Sodium and potassium-sensing macromolecules are involved in the regulation of blood pressure and kidney function. It has been identified that a missense mutation (Glu517Asp) in the calcium sensing receptor is associated with increased sodium sensitivity and hypertension. This mutation was located in a residue that interacts with a third arginine deviating it from its normal interaction with other amino acids at this position in murine homologs. The substitution of aspartate for arginine is thus likely to perturb the three-dimensional structures of these proteins. This may provide new insights into the causes of hypertension, which is characterized by abnormalities in both sodium-induced elevation of blood pressure and sodium retention leading to excessive extracellular fluid volume."

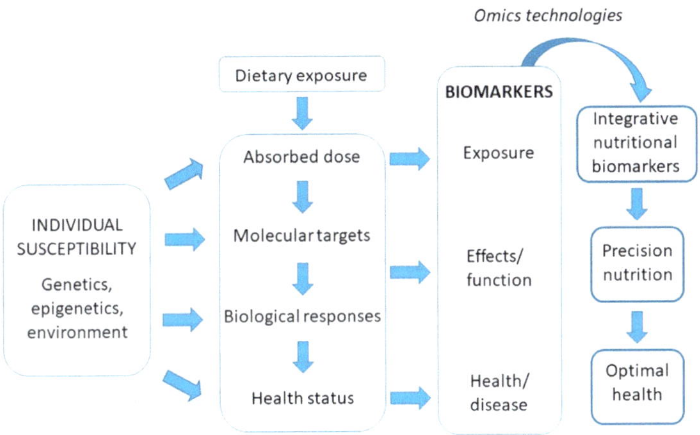

Figure 9.5: The relationship between nutrition and biomarkers helps better understand the mechanisms of health and disease.

Source: Image by MDPI

A study on the relation between nutrition and biomarkers is a way to better understand the underlying mechanisms of health and disease, as well as how dietary inputs affect an individual's risk for developing certain diseases. A randomized controlled trial (RCT) that investigates the effects of specific nutrient intervention on biological markers is called a "NutrIGENomics RCT." The specific pathways and mechanisms affected by nutrition remain unclear and most studies have been observational or cross-sectional; however, prospective studies show that changes in nutritional intake patterns can modify risk factors such as blood pressure, lipids and type 2 diabetes (Piroddi et al., 2017).

The relation between nutrition and biomarkers is a very important factor in the health of a person. A biomarker is "a measurable physiological reaction to an external stimulus or substance that indicates the presence or effect of a biological process, disease or drug treatment." The relation between nutrition and biomarkers is important because when you have a lack of nutrients in your body, it's going to cause an imbalance in your body.

The relation between nutrition and biomarkers is also important because if you're not getting enough nutrients in your body, it can lead to diseases such as cancer and heart disease. A good example of this would be if someone doesn't eat enough fruits or vegetables, then they will not get enough vitamin C in their diet which could possibly lead to scurvy which is caused by not having enough vitamin C in your diet. Another example would be if someone eats too much red meat then they could develop colon cancer because red meat contains more saturated fat than chicken breast does. It's also important because certain foods have been shown to help prevent certain diseases such as cancer so eating those foods regularly might help lower our chances of getting those diseases later on down the road.

Nutrigenomics is a young science, and its influence on nutrition has only just begun. However, it is growing rapidly and creating new research opportunities to supplement the traditional focus of food security and nutritional research. This section provides an overview of the current scientific context for nutrigenomics research and highlights some of the opportunities and challenges facing researchers in this field. In the field of nutrigenomics, there is an entire branch devoted to understanding gene-diet interactions in humans (Bragazzi, 2013). This research uses genomic information to determine how genes interact with diets and how these interactions can be used to prevent and treat disease. Nutrigenomics has been used to develop new treatments for obesity, diabetes, and heart disease.

9.3 CURRENT REGULATIONS SURROUNDING THE PERSONAL GENOME TESTING INCLUDING WHOLE GENOME SEQUENCING AND MICRO-ARRAY ANALYSIS.

9.3.1 The Genetics of Food Preference

Research has shown that genetic variants for food preferences are associated with obesity and may be important targets for future therapy. Genetic variants that influence taste perception may also play a role in determining which foods you like the most. For example, there are genetic differences among individuals in how they process bitter tastes, which leads some people to avoid certain vegetables and other foods. Some of these genetic differences have been linked to more frequent consumption of sweet or fatty foods. No studies have looked at how genetics affect your aversion to bitterness or whether these genes influence how often you eat spicy foods like hot peppers or horseradish root.

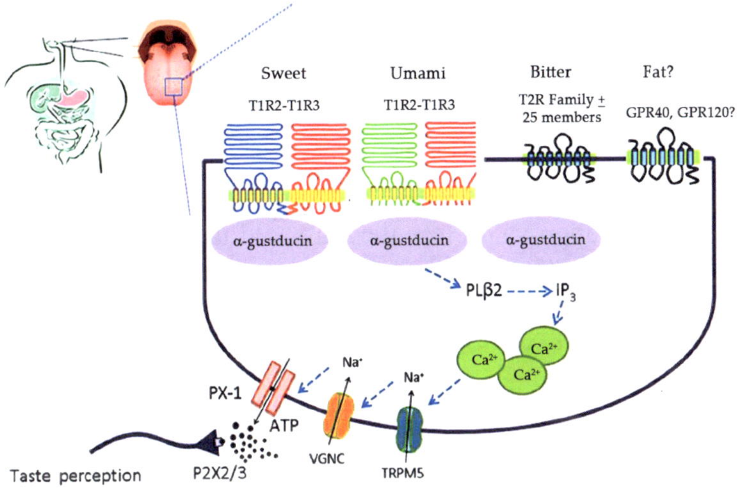

Figure 9.6: Genetic Predisposition.

Source: Lee AA, Owyang C. Sugars, Sweet Taste Receptors, and Brain Responses. Nutrients. 2017; 9(7):653. https://doi.org/10.3390/nu9070653

9.3.2 The Genetics of Food Preparation

This incorporates the genetic differences in;

- Genetic differences in taste perception and preferences
- Genetic differences in food allergies, intolerances, and sensitivities
- Genetic differences in food preparation (i.e., cooking) preferences
- Genetic differences in metabolism of different types of foods

9.3.3 The Genetics of Food Absorption, Storage and Utilization.

The human body absorbs nutrients from food. These nutrients are then stored in fat, muscle and other tissues to be used later. The body uses nutrients to build new cells, make hormones, chemicals and enzymes and produce energy for important bodily functions (Braicu et al., 2017). Nutrients are also used to make new cells that keep us healthy by fighting off germs or viruses that cause diseases like the common cold or flu.

9.4 CONSUMER PROTECTION MECHANISMS

In the absence of a clear regulatory framework, nutrigenomic testing services are largely unregulated. Many companies offer these tests and others use them to promote their products. The FDA has not yet determined the regulatory status of nutrigenomics, so consumers should be aware that there may be no third-party verification or oversight of results or procedures used by these companies.

Consumers should also be aware that many dietary supplements do not require premarket approval before they are sold in the United States. This means that their manufacturers can sell them without demonstrating their safety and effectiveness to FDA. In addition, some laboratories offering nutrigenomic testing may not have been accredited by an agency recognized under U.S. law and may use methods for analyzing DNA samples that lack reliability and validity (the extent to which results reflect actual genetic characteristics).

9.5 RELEVANT POLICY CONSIDERATIONS

Regulatory and policy considerations for nutritional genomics and pharmacogenomics include:

- What are the relevant regulations?
- Are there any regulations in other countries?
- Are there any regulations in other industries?
- Are there any regulations in other fields?

As with any emerging and disruptive technology, it is important for marketers to understand the regulatory status of nutrigenomics. The FDA does not currently regulate nutrigenomics, but as the field gains traction in the medical community and as potential applications emerge for consumers, it will be important to keep abreast of how this rapidly developing science will be regulated (Pal, 2018). There are no clear definitions for what exactly constitutes "nutrigenomics" or even if it should be considered a single field at all; however, current efforts by the FDA indicate that they may eventually consider regulating this burgeoning area of research.

With such a broad scope, nutrigenomics has become something of an umbrella term that covers many different fields including personalized medicine (and its sub-field pharmacogenetics), nutritional genomics (also known as molecular nutrition), synthetic biology/biotechnology and genetic engineering/synthetic biology — all areas where there are many unanswered questions about how they will fit into existing regulatory frameworks.

As science continues to uncover more information about the genetic foundations of human behavior, we may see an increase in the demand for personal genomics testing. When it comes to food and nutrition, this new field of nutrigenomics could have profound implications for our society, ranging from a greater understanding of how humans take in nutrients to increased precision in nutritional recommendations based on individual genetic profiles. Still, there are many questions that need to be answered before nutrigenomics becomes widely available outside of research settings. This is just one example of how policy and regulation can have far-reaching impacts on a host of areas related to health care delivery systems and services.

Regulation and policy considerations have been created based on the rapid advancement of genomic technologies as well as the corresponding issues that arise from them. The concerns surrounding these emerging methods and their effects on individuals, their families and society are considered, as well as their potential impact on future health care options. Current regulations surrounding personal genome testing include whole genome sequencing and microarray analysis (Neeha & Kinth, 2013). Over the past decade, a number of companies have offered personalized medicine testing services, including whole-genome sequencing (WGS), and gene microarray analysis. The ability to test for specific genes gives consumers information on their genetic makeup, allowing them to make better decisions on their health and wellness. However, the general lack of oversight by states and federal agencies has led to concerns regarding data protection and accuracy across these services

Other approaches in modern times incorporate the use of personal genomes, including the emergence of "risk-based regulation" (RBR). RBR is generally considered to be a hybrid model that combines elements of traditional risk-based regulation with elements appropriate to genomics. RBR could facilitate innovation in personalized medicine while continuing to protect patient safety and public health.

9.6 WHOLE GENOME SEQUENCING

Whole-genome sequencing is a revolutionary technology that has the potential to completely change healthcare. It can be used for a plethora of purposes ranging from diagnosing cancer to detecting genetic disorders. Although it is still relatively new and has its drawbacks, whole-genome sequencing is expected to become increasingly popular in the coming years. By 2026, the market for whole-genome sequencing will be valued at approximately $6 billion in the United States alone. This section discusses in detail what exactly whole-genome sequencing is and how this technology works.

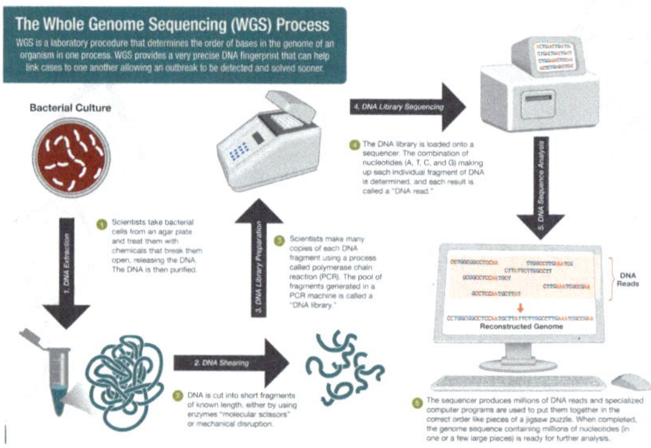

Figure 9.7: Overview of Whole Genome Sequencing.

Source: Image by the Centre for disease control and prevention.

The genomic information of an individual is obtained using various technologies such as "single-nucleotide polymorphism (SNP) microarrays, exome sequencing, and whole-genome sequencing".It is also called genome-wide SNP analysis in which SNPs are identified at different locations on chromosomes. This procedure involves the use of microarray chips containing thousands of SNPs arrayed across chromosomes. The DNA sample is hybridized into the chip and then washed off after staining with a fluorescent dye that binds to DNA(Brennan & de Roos, 2021). The fluorescent signals can be measured by a scanner and analyzed using special software to produce a heat map that shows relative levels of expression for each gene or region on each chromosome.

The cost of whole-genome sequencing has decreased greatly in the last decade from $95 million in 2001 to $1000 to $5000 today. The cost per genome has fallen by a factor of 10,000 or 100 times.

What does it do?

Whole-genome sequencing enables the user to access all the information from the entire human genome in one go. It allows for a comprehensive analysis of genetic variation within a person. Whole-genome sequencing is used to detect genetic variations and mutations, which can be used to diagnose diseases and disorders.

Origin?

Whole-genome sequencing was first performed in 2007 by two companies namely Illumina and Complete Genomics. After the first whole genome was sequenced, there were many more to follow when different companies started offering this service to the public. The first whole genome sequencing was performed by the company named Illumina, which is a California-based company that specializes in technologies for DNA sequencing and analysis. Their services include microarray-based genotyping and gene expression profiling, as well as DNA sequencing at high throughput levels using their own technology known as Sanger Sequencing (Illumina website).

Complete Genomics Inc., another California-based company also offers services related to genomics including whole genome sequencing, targeted resequencing services for both human genomes and exomes (gene regions), on-demand diagnostic testing from blood samples or urine samples from newborn babies who have no symptoms of illness but require further evaluation. The number of individuals who will be sequenced is expected to increase from 1 million in 2015 to more than 10 million by 2022 (Bionaz et al., 2015). The number of whole-genome sequencing companies is also expected to grow as the demand for this type of testing increases. It is predicted that by 2026, the market for whole-genome sequencing will be valued at approximately $6 billion.

Applications

Whole genome sequencing can be used for a variety of applications such as cancer genomics, reproductive health, pharmacogenomics, forensic genomics, etc. The following is a list of some potential uses:

- Detecting genetic cancer predisposition
- Advancing clinical care through genetic testing
- Understanding population history and diversity

Whole-genome sequencing has the potential to revolutionize healthcare and diagnosis in the future. Whole-genome sequencing can be used for a variety of applications such as cancer genomics, reproductive health, pharmacogenomics, forensic genomics, etc. While there are still many challenges that need to be addressed before this technology can be used in healthcare, it has the potential to revolutionize many areas of healthcare and disease diagnosis. The development of cost-effective whole genome sequencing has already resulted in major improvements in the field of cancer genomics.

9.7 MICRO ARRAY ANALYSIS

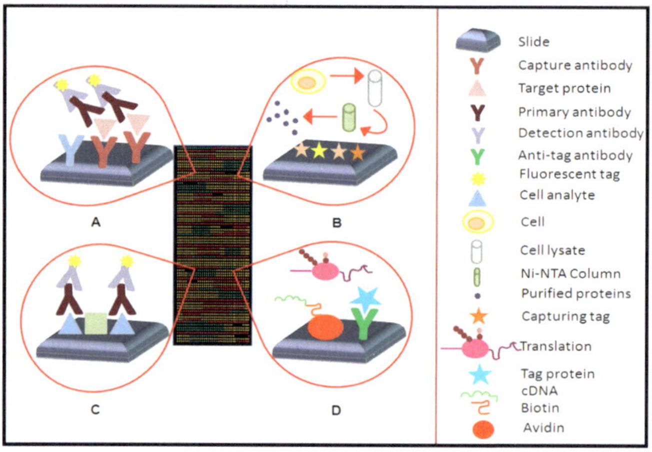

Figure 9.8: The DNA Micro-Array analysis.

Source: Díez P, Dasilva N, González-González M, Matarraz S, Casado-Vela J, Orfao A, Fuentes M. Data Analysis Strategies for Protein Microarrays. Microarrays. 2012; 1(2):64-83. https://doi.org/10.3390/microarrays1020064

Microarray analysis is a process used to determine the expression levels of genes in a particular array. This type of analysis is used in nutrigenomics to determine which genes are expressed under certain conditions, such as when a person is exposed to different foods. In other words, microarray analysis can be described as the process of identifying the different types of genetic markers within a given genome. The term nutrigenomics was coined by Dr. Michael Gershon and Dr. James Greenblatt in 2004 to describe the study of how the food we eat affects our genes, and how those effects can be passed down to future generations. It uses microarrays to detect and quantify the presence of specific DNA, RNA, or protein molecules in a sample (McNamara, 2010). This technique allows for the detection of thousands of different gene expressions at once, and it's been used in many different fields, including nutrigenomics.

Microarray analysis is a powerful tool for identifying gene expression patterns in cells. The technique has been used to identify many new biomarkers and potential therapeutic targets, as well as to study the effects of disease and drug treatment on gene expression. However, microarray analysis is only as good as the quality of its data.

The accuracy of microarray results can be affected by many factors, including sample quality, experimental design, and data processing. For example, one major source of error is cross-hybridization between probes on different arrays. Probes that are related by sequence similarity can compete with each other at the hybridization step, resulting in an underestimation of true gene expression levels or even erroneous detection of false positives (1). Similarly, probes that are too close together on the array (2) or have imperfect matches to their target sequences (3) may also lead to inaccurate results.

Another major challenge faced by microarray users is how best to analyze data from multiple experiments or replicate experiments conducted on different platforms (4). This problem is exacerbated by the fact that most microarray data sets contain a small number of samples per condition often fewer than five which makes statistical analysis easier. Another challenge is that microarray experiments are often expensive and require a lot of equipment (Mustapa et al., 2020). This can make it difficult for smaller labs, particularly those located outside major cities to access this kind of top-notch technology without spending a lot of money on their own equipment or renting time on someone else's machines.

Another major challenge faced by microarray users is the fact that there are so many different kinds of cells and tissues, each with its own unique properties. In addition, there are many different ways to measure gene expression, from fluorescent probes to RNA sequencing. Each method has its own strengths and weaknesses, making it difficult to compare data across studies. Also, given the fact that these arrays can detect expression patterns of thousands of genes at once, there's no way for humans to interpret what those patterns mean or how they should be interpreted.

9.8 RISK-BASED REGULATIONS

As nutrigenomics becomes more common, regulatory bodies around the world are looking for ways to regulate it. One of the main approaches that are being taken is risk-based regulation: regulating products based on the amount of risk they pose to consumers.

Risk-based regulations are typically considered in two contexts: pharmaceuticals and food. In the case of pharmaceuticals, there is a long history of risk-based regulation in place. In the US, drugs are deemed safe enough to be used if they meet certain criteria regarding their impact on patient health, including how likely they are to cause side effects or serious harm.

For food products, however, there has been less regulation around risk assessment until recently—but now governments are starting to pay attention. For example, in Europe where nutrigenomics is much more common than it is in America or Canada, there has been more focus on applying risk-based assessments for new foods like genetically modified organisms (GMOs).

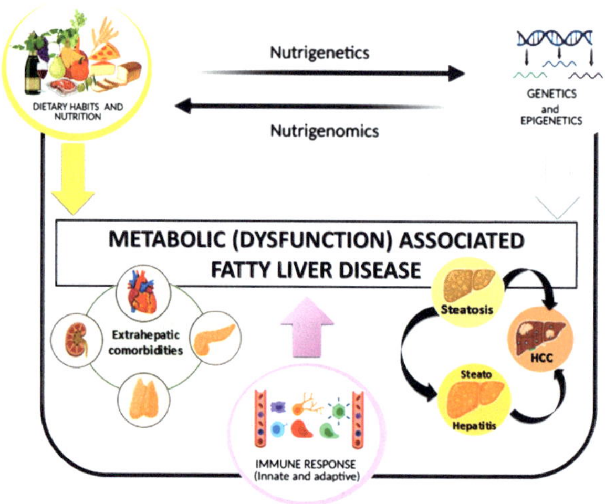

Figure 9.9: An example of the risk-based regulation.

Source: https://www.mdpi.com/2072-6643/13/5/1679

While this type of regulation can be applied to any industry or sector, it is most commonly used in the pharmaceutical and nutrition industries due to its impact on public health. In order to determine whether a specific activity should be regulated based on risk, regulators consider both the potential benefits and risks associated with the activity in question. If the benefits outweigh the risks, then no additional regulation is needed (Masotti et al., 2010). However, if there are significant risks involved in an activity for which there are no apparent benefits, then it may be subject to more stringent regulation.

The field of nutrigenomics is in most cases misunderstood and has been overlooked in regulatory policy for decades. As such, there are few risk-based regulations in place to protect consumers from companies that use nutrigenomic information in their marketing or advertising. This is a problem because nutrigenomics can be used to target specific populations with misleading health claims about products or services that may not be beneficial for them.

Additionally, there are many questions about how long this information stays around in our bodies after we consume it. For example, if I eat a certain food today and then decide later that I don't like it anymore, will my body still have traces of it six months later? Will those traces still reveal what foods I've eaten over time? These questions need answers so that consumers can feel confident using nutrigenomics as part of their daily lives without worrying about what information might be revealed down the line.

The growing field of nutrigenomics is leading to new regulations around the world. While some countries are taking a risk-based approach to regulation, others are choosing a more strict and rigid system of rules.

9.9 THE FEDERAL CLINICAL-LABORATORY IMPROVEMENT AMENDMENTS

The Federal Clinical-Laboratory improvement amendments are a set of federal rules and laws that govern clinical laboratories and their operations. These regulations were passed in order to ensure that patients have access to safe, effective healthcare services. The amendments define quality standards for both diagnosis and treatment of disease, as well as provide guidelines for reporting adverse events related to lab testing errors. The federal Clinical Laboratory Improvement Amendments (CLIA) regulate all laboratory testing (except research) performed on humans in the U.S., establishing quality standards for the accuracy, reliability and timeliness of patient test results regardless of where the test was performed, who performed it or what technology was used (Castro et al., 2006).

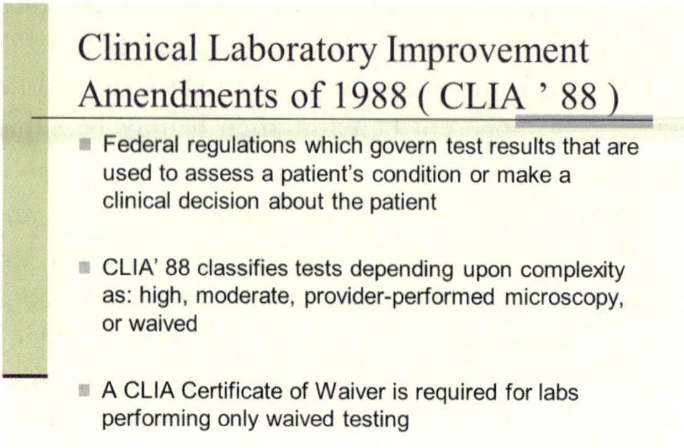

Figure 9.10: CLIA` 88.

Source: Image by Slide player

Background

In 2015, President Obama signed into law an update to Title IV of LHDPA (Laboratory Health & Safety). This legislation primarily focuses on improving patient safety through increased oversight by state medical boards over clinical labs across America – including those accredited by CAP or COLA – while also creating greater transparency when it comes down to how doctors use test results provided by these facilities."

9.10 HOW IS THIS AMENDMENT VIEWED AS A POLICY CONSIDERATION ON NUTRIGENOMICS?

The CLIA has updated their regulations to include the term "nutrigenomics" in their policies and procedures. This is important because it means they are recognizing this emerging field of study, with its potential to improve patient care and save lives by enabling early diagnosis and intervention of disease. The Federal Clinical Laboratory Improvement Amendments (CLIA) regulate all laboratory testing (except research) performed on humans in the U.S., establishing quality standards for the accuracy, reliability and timeliness of patient test results regardless of where the test was performed, who performed it or what technology was used.

CLIA was passed in 1988 to protect the public from inaccurate and unreliable health care results. CLIA specifies how laboratories must operate, and it applies to all clinical laboratory testing (except research) performed on humans in the United States. Laboratories that test human specimens for microbiology, chemistry, hematology or immunology use; blood gas analysis; coagulation tests; urine analysis; or any other specimen that is collected from a patient primarily for diagnostic purposes are subject to CLIA regulations (Carlberg, 2019).

Based on the policy considerations outlined in this review, it is evident that there are many challenges facing genomics and nutrigenomics as they move from a research focus to clinical use. A variety of different guidelines and policies exist across the U.S. that regulate specific aspects of genetic testing or dietary supplements, but there is no overarching regulatory framework for nutrigenomic tests. Challenges related to nutrigenomic tests include lack of sensitivity, specificity, reproducibility and validation; the potential for false-positive and false-negative results; test complexity; interpretation issues; cost implications; ethical concerns with confidentiality and data use; genetic discrimination (e.g., in health insurance coverage or employment); lack of reimbursement by insurers for personalized medicine approaches to nutrition care; administrative barriers such as cumbersome credentialing processes needed by some insurers before a provider can order tests from specific laboratories (e.g., Medicare's Therapeutic Pathway program). These challenges are compounded by limited public understanding of genetics and nutrient interactions, including consumers' expectations about how much control they have over their gene expression through diet.

9.11 THE CENTER FOR MEDICARE AND MEDICAID SERVICES (CMS) REGULATIONS

This is covered by the Patient Centered Outcomes Research Institute (PCORI). The Patient Centered Outcomes Research Institute (PCORI) is a nonprofit organization that was created by Congress in 2010 to expand patient-centered outcomes research. PCORI is a public-private partnership funded by the federal government, pharmaceutical industry, medical device industry, health insurance industry, and the medical profession.

CMS

CENTERS FOR MEDICARE & MEDICAID SERVICES

Figure 9.11: The CMS agency is in charge of regulations and policies that relate to Nutrigenomics.

Source: Image by Policy and Medicine

Coverage with Evidence Development (CED) is a mechanism to cover new and innovative therapies, drugs, medical devices or diagnostic technologies that lack the necessary evidence of safety and effectiveness. In order for an applicant to receive CED coverage from CMS, they must submit all the necessary information (including clinical data) so that CMS can determine whether the proposed use meets their criteria for CED. If it does, then CMS will make arrangements with the applicant regarding how much will be paid out-of-pocket by patients who are enrolled in this program (Mathers, 2017). This amount can vary depending on whether it's an experimental drug or device being tested; if it works well enough during Phase II trials but has some side effects after Phase III trials begin testing larger populations; or if there's little chance of success but still uses up valuable resources without offering any benefit at all.

In addition to the various regulations that govern Medicare and Medicaid, individual providers are also required to adhere to national coverage determinations (NCDs). An NCD is essentially a way for CMS to clarify whether or not it will cover certain services. For example, if a healthcare provider has an NCD on their practice site that states they can't provide a specific service, then they must inform their patients of this by posting it somewhere on the premises. The purpose of these determinations is twofold: firstly, they help ensure that patients aren't receiving unnecessary treatment; secondly and more importantly, they protect providers from any potential liability issues related to performing services outside of what's been approved by CMS/Medicare/Medicaid. MS has several mechanisms in place for determining when an NCD should be issued; however, there are some limitations as well (which we'll get into later). By gaining access to our database containing thousands of these documents we've been able to make it easier than ever before for you!

As it is known, nutrigenomics is a new science. The Center for Medicare and Medicaid Services (CMS) has recently issued regulatory changes that are changing the way we research, treat patients, and understand disease and nutrition. These regulations will require all healthcare providers to be trained in this new field of knowledge in order to receive reimbursement from CMS programs. It also requires that genetic tests performed on patients be conducted by qualified individuals who have been trained and certified as well as documented proof of these certifications maintained by each healthcare provider's practice management system (Loor et al., 2015). These new requirements may require some changes in how your organization handles genetic testing services or that you work with a partner who can help you meet these new requirements directly or through outsourcing arrangements such as cloud-based solutions

We could see the need for more research in this area after reviewing current regulations and policies. Hopefully, we've demonstrated that there are many ways for CMS or other agencies like PCORI to support research into nutrigenomics—and it's time for all stakeholders involved to get on board. The potential benefits from this work would be huge, both for the health of our country and its budget.

9.12 ROLE OF THE SCIENTIFIC ADVISORY COMMITTEE ON NUTRITION

The Scientific Advisory Committee on Nutrition (SACN) has for some years had an interest in the role of genetics in nutrition, and the impact of diet on disease risk. The SACN is an independent scientific advisory committee that provides advice to the government on the relationship between nutrition and health. The purpose of its reports is to provide evidence-based advice on specific issues related to nutrition, so that policymakers can make informed decisions about the advice they give to the public.

sacn

Scientific Advisory Committee on Nutrition

Figure 9.12: The SACN advises governmental bodies on regulations that relate to nutrigenomics.

Source: Image by Twitter

In particular, SACN published a report on diet and cancer in 2004 which considered various aspects of diet and cancer risk including the role of genetics. In following this up, SACN published a further report entitled 'Nutrition & Health' in 2006 which provided an overview of key issues relating to nutrition; including genetic variation as it might affect nutritional requirements or disease susceptibility.

Since then, there has been much discussion about how genomics might be used to help understand why people respond differently to nutrients or foods. SACN commissioned Professor Simon Capewell - a member of its Working Group on Nutritional Epidemiology -to undertake research into this area. The results were published in two papers: one paper was focused specifically on looking at how genetic variants could be used as markers for dietary intake, while the second paper assessed whether there was evidence that these variants are associated with disease outcomes (Chen & Zhao, 2013).

In February 2016 SACN published an update to its 2009 report on Carbohydrates and Health, which provided recommendations regarding carbohydrates as part of a healthy diet. The scientific evidence base underpinning nutrigenomics has grown over recent years but it remains limited compared with other areas in nutritional science such as genomics or metabolomics. There is no clear consensus on how best this emerging field should be regulated or what measures are likely (or not) to achieve the intended outcomes of regulation: I) protecting vulnerable consumers from misleading claims; ii) preserving consumer choice; iii) enabling responsible innovation by industry; iv) providing clarity around labelling requirements for foods and supplements incorporating nutrigenetic information; v) ensuring appropriate communication between industry/regulatory bodies/the public etc., including scientists conducting research into nutrigenetics-related products and services; vi) fostering trust in this new field by consumers who may be unfamiliar with it at present but could become interested once more information becomes available through media coverage etc.; vii) minimizing unnecessary costs associated with implementing new regulations where existing laws already provide sufficient protection against misleading/unfair practices such as false advertising under Consumer Protection Regulations 2008; viii) maintaining consistency across countries where similar legislation exists or is being developed (e.g., Australia).

SACN received a submission from a member of the public in response to its consultation on carbohydrates and health. The company concerned, Vitl Life (formerly Pro-activ), proposed to undertake personalized nutrition based on nutrigenomics, which would include providing dietary supplements containing phytochemicals and/or probiotics. Pro-activ (subsequently Vitl Life) sought approval for nutrigenomic tests to determine a person's susceptibility to various diseases and how food affects their health (Ladeira et al., 2018). Nutrigenomics is the study of how food can affect gene expression, which in turn influences our health outcomes. The response from Health Canada was that it would not regulate these tests because they did not pose any immediate safety concerns, despite the fact that there were no guidelines or standards in place with respect to testing, reporting or interpretation of results.

After this decision was made public, several groups expressed concern over what they viewed as insufficient regulation and over privacy concerns associated with genetic information being made available to third parties without consent or knowledge. They also expressed concern that consumers may be misled into thinking that "personalized nutrition" based on genetics will lead to improved health outcomes when most current research indicates otherwise.

Indeed, Nutrigenomics is a relatively new field that aims to use information about individual genetic variation in order to inform dietary habits and improve health outcomes. This paper describes how it can be used in practice by looking at areas such as: 1) defining nutrigenomic traits; 2) understanding how genetics affect nutrient metabolism; 3) examining current knowledge on gene-diet interactions associated with diseases including cardiovascular disease (CVD), type 2 diabetes mellitus (T2DM), obesity, inflammatory bowel disease (IBD), rheumatoid arthritis (RA) and cancer; 4) exploring potential applications of personalized nutrition within primary care settings; 5) considering regulatory issues surrounding claims made around nutritional genomics products or services; 6). Identifying recommendations for practitioners/professionals working across disciplines including dietetics/nutritionists, and healthcare professionals providing advice about lifestyle changes/habits such as physical activity levels or smoking cessation strategies - these individuals need additional training so that they can adopt practices informed by current evidence about their role in improving health outcomes through nutritional interventions(Kussmann & Affolter, 2009).

9.13 NUTRIGENOMIC TESTING

This explores the relationship between nutrition and genes and how an individual's genes affect their response to certain foods. The emerging role of nutrigenomics has brought new possibilities for personalized dietary recommendations and its application in clinical practice. It highlights the importance of setting up a regulatory framework for different stakeholders involved with nutrigenomics testing, including nutrition professionals and organizations providing such tests to consumers. Nutrigenomic testing involves analyzing DNA methylation patterns through multiple technologies (e.g., microarray analysis) to determine whether certain foods affect gene expression by changing epigenetic marks.

In order to understand the role that nutrigenomic testing can play in clinical practice, it is important to first understand what the technology entails. In general terms, nutrigenomic testing is a type of genetic testing that assesses the effect of nutrition on the genome (1). The term "nutrigenomics" was originally used by Stephen Rothberg and his colleagues at Genetic Systems Corporation as early as 1986 (2) and refers to how dietary components influence gene expression. This concept was further developed by researchers at Stanford University who used microarray technology for genome-wide profiling studies on mice fed with different diets (3). Nutrigenomic tests use genomic information obtained from blood samples collected after fasting for 12-14 hours before sample collection; DNA extracted from these samples are then analyzed using microarray chips containing thousands of genes known or suspected to be involved in disease processes (4).

Nutrigenomic testing is not regulated by the FDA. In the United States, nutrigenomic testing is regulated by the Food and Drug Administration

(FDA). In order for a company to market a product as "nutrigenomics," it must be approved by the FDA. If a company does not have such approval or if its products are not marketed as being "nutrigenomics" as defined by FDA guidelines, then those products cannot be considered to be nutrigenomic in nature (Corella & Ordovas, 2009).

The FDA does not regulate companies that sell nutritional supplements or dietary supplements in any way—including whether these products are labeled as nutraceuticals, nutraceuticals-like substances, or genetically modified foods (GMOs) that provide some health benefit when consumed regularly over time; nor does it regulate their use outside of what has been approved by their manufacturer(s). The only way for someone within this country who wishes to use one kind of nutritional supplement/dietary supplement product over another would be if they could afford to do so with no assistance from either private insurance companies or government assistance programs like Medicare Part B premiums paid out annually during tax season before April 15th each year (which will help cover costs associated with having enough money saved up before making purchases made throughout late winter months).

In the United States and other countries, consumer protection regulations are being developed to protect consumers from misleading claims or deceptive practices by companies that offer nutrigenomic tests. The Federal Trade Commission (FTC) is responsible for enforcing these laws.

In December 2014, the FTC issued a report on its investigation into the marketing of personal genome sequencing services. The report identified several areas where genetic testing companies made claims that were unsubstantiated or misleading, including:

- Promises of personalized health advice based on DNA analysis (such as a cancer risk assessment) without providing any supporting scientific evidence;
- Claims about links between genetic variants and diseases when no link has been established by peer-reviewed studies; and
- Unsubstantiated claims about gene-diet interactions affecting health conditions such as diabetes or obesity (Connolly & Dawson, 2015).

Nutrigenomics is a new field of research. It is important to regulate the marketing and advertising of products and services incorporating nutrigenomics. Recommendations and guidelines for the conduct of research are also needed.

CHAPTER 10

REGULATION AND POLICY CONSIDERATIONS OF NUTRIGENOMICS

CONTENTS

10.1 INTRODUCTION

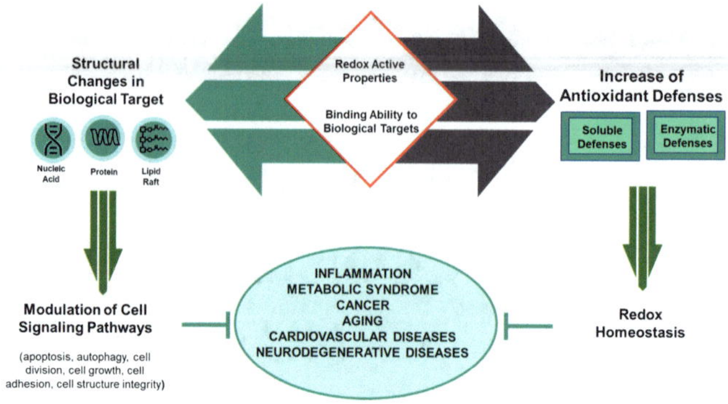

Figure 10.1: Nutrigenomics Caloric Restriction.

Source: Caradonna F, Consiglio O, Luparello C, Gentile C. Science and Healthy Meals in the World: Nutritional Epigenomics and Nutrigenetics of the Mediterranean Diet. Nutrients. 2020; 12(6):1748. https://doi.org/10.3390/nu12061748

The bulk of the scientific and economic revolutions that have taken place may be linked back to advancements in technology. Because this new equipment is now readily available, researchers can take advantage of contemporary technology to get a fresh viewpoint on the natural world. To put it another way, when these ideas are formalized and made part of the non-scientific public's common knowledge, they provide the foundation for commercial businesses to exist and develop. Because of this, improvements in the economic world and the sciences are intricately interwoven, and both are helped by advancements in technology. Every day, new technical and scientific discoveries are created; yet, only a tiny fraction of them will culminate in a scientific or commercial revolution (Kussmann & Van Bladeren, 2011). This is because scientific and commercial revolutions require a critical mass of participants. Antibiotics, the internet, wireless telephony, polymerase chain reaction (PCR), microchip arrays, freeze-drying technologies for food, and round-ready soybeans are examples of the sorts of breakthroughs that fall within this category.

Some examples of technologies that did not have much of an influence or none at all include cold fusion, vitamin O, and mega dosages of vitamin C. To accomplish this, you need to have an understanding of the technologies, the quantity of research that is presently being carried out, and the industries that are anticipated to be impacted. There is no consistent pattern that can be utilized as a basis for anticipating these changes in legislative policy. These kinds of legislative shifts have the potential to usher in whole new tendencies or have a big influence on ones that are already prevalent. It is hard to anticipate which of the many technologies will have an effect on the environment, and merely predicting change is not enough to address the problem. Businesses will want further knowledge on the topic before they can formulate a response to the launch of new technology (Kang, 2012). Understanding how new technology will function, how it will impact a business's operations, and how it may be incorporated into the broader strategy of the organization are all key components of a firm's capacity to adapt to changes in the way technology is utilized.

A management team that predicts how the new environment will affect their capacity to reach consumers and give value to them will have an easier time swiftly adapting to the new environment. Those that can predict changes and adapt correctly to them are more likely to be successful than companies that merely react to them when they occur. The Long-Term Ecological Research Organization (LSRO) and the Center for Emerging Science Issues collaborated to form this organization. The senior executives in both the public and commercial sectors can get scientific evaluations and suggestions from the LSRO, which is a nonprofit organization. For over 40 years, LSRO has been organizing meetings of expert panels to investigate and report on the most recent findings of the state of scientific knowledge in a broad variety of biological, clinical, agricultural, and environmental sectors (Irimie et al., 2019). These committees do research and produce reports on the condition of the scientific community. The Center for Emerging Topics in Science (CEIS) was established by LSRO with the goals of identifying emerging challenges in research and technology that have the potential to change the landscape of the field and investigating the potential effects that these challenges may have on science and industry. To assist the industry in anticipating and adjusting to shifting conditions, LSRO CEIS will draw on LSRO's significant expertise in examining the state of research at present.

10.2 NUTRIGENOMICS FUTURE

Figure 10.2: Nutrigenomics Future.

Source: https://kajabi-storefronts-production.kajabi-cdn.com/kajabi-storefronts-production/blogs/22934/images/asKtImSpWOlX2LlX07w5_BHH_-_Blog_Nutrigenomics.jpg

CEIS was tasked with investigating the effects of the genomics and information technology revolution, with a particular focus on the rapidly developing field of nutrigenomics. How food and supplement bioactive chemicals can affect gene expression or structure is the focus of Nutrigenomics (also known as nutrigenomics). The discovery of the essential nutrients and the deficiencies that go along with them was one of the most important achievements of the twentieth century. It is possible to make a more specific diagnosis and general recommendations because of this. Nutritional advice and recommendations for public health are more likely to be universally applicable when they are based on statistical analysis (Hurlimann et al., 2014). When it comes to evaluating success, factors like the decreased prevalence of deficiency diseases or an increase in life expectancy can be used to measure the process's success. Limiting the benefits of nutrition only to the avoidance of nutritional deficits, on the other hand, is an overly narrow view of what eating can do for our health. It's crystal clear that no solution exists that can be used in all circumstances.

First-generation solutions like the food guide pyramid are no longer sufficient, and a second-generation has emerged. In the second generation of solutions, adapted pyramids are available for specific age groups, ethnicities, vegetarians, and other groups. Even though these programs are a good start, they fail to take full advantage of the progress made in biomedical science. As a result, this discovery has come at a time when rapid advances in knowledge, particularly in the fields of genomics and information technology, are taking place. Proteome (a collection of proteins encoded in the genome), metabolome (a collection of metabolic entities produced in cells, tissues, organs, and species in various states of health or disease), epigenome (a collection of DNA modifications that alter genome function but do not alter the DNA sequence) and other post-translational modifications that influence gene expression have all been the subject of intense investigation (Hesketh, 2013). To truly meet the challenges of the twenty-first century, it will be necessary to go beyond simply creating diagnostics and medications to try and treat diseases that have already manifested, and instead integrate this scientific understanding and provide personalized recommendations for diet, lifestyle, and drug treatment to help people stay healthy and avoid disease. Individual people, businesses, and society as a whole will all benefit from this information.

A better understanding of how genes and the environment interact to affect a person's health and performance will be gained as more data is accumulated. Through this research, we hope to learn more about how gene expression influences an individual's reactions as well as how things like nutrition and the environment affect gene expression. As a result, nutrigenomics-based personalization of medical care and dietary intake is no longer an option. Increased awareness of the possibility that dietary changes or the foods one consumes can improve one's health or reduce the risk of dietary-related disorders is fueling the excitement surrounding nutrigenomics. Individual genetic predispositions for chronic illnesses and the ability of each person's body to respond to dietary intervention may now be accurately diagnosed thanks to this kind of analysis.

By taking the right number of nutrients, many disorders can be easily averted. Bioinformatics, nutrition, molecular biology, genomics, epidemiology, and molecular medicine are all included in the study of nutrigenomics. In addition to the study of genomes, "nutrigenomics" includes the study of epigenomics, post-translational modifications, proteomics, and metabolomics (Gonzalez et al., 2015). To explore the challenges that this new field of research would bring, especially in the areas of individualized nutrition and diets, CEIS arranged a panel of experts from the disciplines of science, technology, and industry.

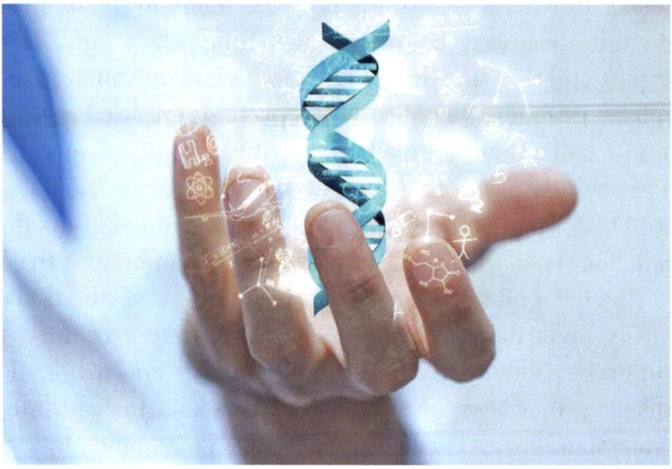

Figure 10.3: Nutrigenomics will shape the future of food and nutrition.

Source: https://nutrigenomicsinstitute.com/wp-content/uploads/2018/07/TestsNutrigen%C3%A9ticos.jpg

Assembling has started for the instruments needed to generate an individual's genomic, epigenomic, proteomic, and metabolomic profiles. Using these profiles, we can determine whether or not a medication will be effective, what dosages may be harmful, whether or not they should adhere to a low-carbohydrate diet, whether or not standard nutritional counseling will be beneficial, or counterproductive, and whether or not they have specific nutrient requirements. With this knowledge, medical providers will be better equipped to give tailored recommendations while avoiding the hazards associated with population-based theories (Guéant et al., 2014). This will bring health-promoting advice down to the individual level, going far further than the present individualized dietary counseling pyramid, which is still based on notions relevant to the entire community. Internet-based firms are presently giving clients personalized DNA profiles and suggestions that are tailored to the customer's specific profile.

Even if the current iteration of the services may consist of more hype than the actual value, they nevertheless have the potential to change the industry provided they are supplied acceptably by dependable and competent firms and are established on sound scientific data. This vision of a tailored and health-promoting future was adopted by the expert panel, and they proceeded to identify and prioritize several areas of nutrition that are ripe for this sort of approach. They also devised a multi-partner model that will allow the industry to take advantage of this strategy, and they outlined the hurdles that need to be solved for it to become a reality. After studying the research that revealed how individuals react differently to dietary modifications, the panel of industry experts concluded that we are on the edge of a commercial revolution (Fenech et al., 2011). The following is a sampling of the judgments and recommendations provided by the expert panel: As long as clients think they have some degree of influence over the information, it is already evident that they want information that is suited to their requirements. Individualized nutrition advice and products will probably definitely be embraced first by the "worried, affluent well," with other customer segments following later.

These sorts of firms that can make those personalized delicacies at the same time will be the industrial giants of the future. This product's profit margins will be much larger than those of ordinary functional meals. Nutritional intervention during the earliest potential moments in the life cycle, such as during pre-pregnancy, pregnancy, early infancy, and so on, has been proved to have considerable long-term health advantages. These advantages include a lower risk of chronic illnesses such as diabetes and cardiovascular disease. However, the implications of life-long impacts of dietary choices, particularly early in life, might encourage quick adoption of dietary choices that are a reflection of the heightened nutritional vulnerability of early infancy in some parts of the consumer population. The future rests in epigenetics and other post-translational processes. It won't be essential for firms to make hundreds of various sorts of things anymore, as that's not what buyers want. The panel highlighted connections between a company's planned approach of providing personalized healthy foods and a shoe store. The mass market is supposed to manufacture and sell a restricted number of stock products that are 'adapted' to varied client desires and tastes. This is in contrast to the luxury sector, which can make shoes for each unique consumer.

As medical research uncovers new nutrigenomic capabilities and manufacturing developments combine to deliver a wider range of goods and, as a result, the best possible match for the requirements of individual customers, the total number of products and the variety of those products may, in the end, grow and become more varied. A comprehensive diagnostics component is important for individualized nutrition to be adopted. It is presently accessible to purchase diagnostic equipment; however, this technology needs to be developed for it to effectively evaluate the health state of healthy persons and, as a consequence, become more user-friendly

(Ferguson et al., 2010). The provision of consumer feedback on efficacy, in addition to an awareness of the requirements of each individual, will make diagnostics a crucial component. There have been several unsuccessful attempts at illness prevention and health promotion that were designed to offer positive reinforcement and a sense of accomplishment. In its absence, compliance and consumer interest will, over time, progressively erode. These brand-new personalized dietary directions will almost definitely be offered by a health care practitioner (HCP) or another trustworthy source. In addition to offering specialized dietary advice and monitoring remarks, the HCP will aid in the interpretation of the findings of the diagnostic tests.

This health care practitioner is not likely to be a medical doctor; nevertheless, the function may be filled by pharmacists, nurses, dietitians, or other competent paraprofessionals. For instance, Pharmica has begun selling healthful meals and supplements while at the same time teaching pharmacists to become advisers. It is likely that health maintenance organizations, often known as HMOs, may be employed as HCPs and may even provide the items as one of the services they provide. HMOs may opt to include this into their continuous health promotion activities to attract the concerned and rich in addition to consumers. · A tailored diet that is grounded on sound research is the only way to assure its effectiveness (Ferguson, 2016). Customers have become increasingly cautious regarding the authenticity of product claims and the efficacy of dietary supplements and health foods as a result of competition in the health food and dietary supplement sectors. A new business paradigm that is based on verifiable personal evaluation, such as the one that is envisioned below, would not be based on promise, but rather on genuine personal demonstrations of efficacy, and as a result, it will not be particularly susceptible to bad news or the reaction of consumers.

A large amount of bioinformatics will need to be included for individualized nutrition to be achievable. The great majority of these technical improvements are now accessible, but for them to realize their full potential, considerable financial investment in research and development is necessary. It is now beyond the capability of the information technology that is available to undertake an examination of a person's full genome or metabolome. In addition, the multiple single nucleotide polymorphisms (SNP), metabolic markers, and other similar variables have not yet been fully explored or related to an individual's health state or the risk of disease. Nevertheless, there are bits of the data that are accessible and may be instantaneously turned into concepts with significance. We are almost completed incorporating valuable personalized advice for health promotion.

Further research and refinement will entail the use of new markers, the integration of more information, and the discovery of additional potentially sensitive health outcomes. This personalized diet plan may not work instantly with all commodities because of how precise it is. Acceptance from consumers shortly will most likely be dependent on making the proper selection of items that have the "permission" of customers to be customized orange juice, probiotics drinks, etc).

There are some health and physiologic goals for which there are not presently adequate data to establish whether or not they will benefit from the new paradigm or approach to health care. The health outcomes and bioactive compounds were under review and given attention by the expert group. In terms of health and physiologic results, illnesses such as cardiovascular disease, cancer (particularly prostate, colon, and breast), type II diabetes, taste acuity, and inflammation are in the greatest position to take off. Following closely after are diseases such as lactose intolerance, obesity, decreased taste perception, immunological disorders, and allergic responses. Dementia, Alzheimer's disease, other memory, and cognitive difficulties, macular degeneration and cataracts, osteoarthritis, muscle atrophy, and gut function are some of the additional illnesses that deserve inquiry for the goal of enhancing health through nutrigenomics (Fenech, 2008). There is not going to be a single firm or industry that will be able to market its nutrigenomic goods on its own. To be successful in the marketplace, you will need to collaborate and communicate across a broad variety of industrial segments, even if it is uncertain how these segments will interact with one another. The expert group provided a variety of possible implementation techniques.

Participation from food producers, food manufacturers, makers of dietary supplements, diagnostic firms, health care providers, HMOs or other health insurance groups, and venture capitalists or other investors are all examples of these sorts of operations. In a similar vein, scientific gains in basic knowledge will demand the collaboration of scientists with a wide variety of skills as well as the development of scientific consortiums to mature these unique notions into an established area of research.

10.3 FOUR IMPLICATIONS FOR THE FUTURE

Nutrigenomics has enormous promise as a scientific topic of study. In order for nutrigenomics to attain its full potential, there are a number of issues that need to be addressed outside of science. Ethics, economics, business, and public policy are only a few examples.

10.3.1 Public Attitudes towards Nutrigenomics

Figure 10.4: Public Attitudes towards Nutrigenomics.

Source: Malsagova KA, Kopylov AT, Sinitsyna AA, Stepanov AA, Izotov AA, Butkova TV, Chingin K, Klyuchnikov MS, Kaysheva AL. Sports Nutrition: Diets, Selection Factors, Recommendations. Nutrients. 2021; 13(11):3771. https://doi.org/10.3390/nu13113771

The general public's perception of nutrigenomics, as David Castle points out, will have a significant impact on the field's acceptance. Castle's research in commodities and services focuses on how new technologies such as nutrigenomics succeed or fail. He also investigates how scientific and technological agendas are developed. The public's attitude toward technology is one of the most important factors. People are worried about nutrition genomics for a variety of reasons, all of which are valid. Is there any evidence that nutrigenomics, for example, can improve an individual's health? When following a nutrigenomics-based nutritional prescription, there must be a balance of benefits and risks, as well as an understanding of the causal relationships being investigated. Obtaining data for the creation of a dietary prescription frequently causes concern about what happens to genetic information provided to a company or clinic for testing.

Customers should be aware of who has access to their data and whether or not their employers or insurance companies have access to it. There have been complaints about how a public database is used because it contains personal information. Are you going to burn it or leave it somewhere for others to find if they want to? Consumers are also concerned about the regulation of nutrigenomics technology. Castle stated that they want the examinations to be regulated. "They want health claims regulated," said the source. Consumers, for their part, recognize that nutrigenomics will necessitate a different set of regulations than those currently in place for drugs and food and that it may even combine elements of both. "The major challenge for regulators," according to Castle, "is defining what new inventive regulatory frameworks must be established to handle this sophisticated business." Castle's conclusion is, "Because the government is accustomed to regulating within vertically well-organized, highly connected domains such as drugs or food, developing new guidelines for nutrigenomics is difficult (Fallaize et al., 2013)". There are also concerns about access and equity. There is widespread concern that nutrigenomics will be limited to a wealthy few, leaving out the vast majority of the world's population.

These concerns are intertwined with the problem of effectively communicating nutrigenomics to the general public. Castle has made significant contributions as a pioneer in the field of private and public delivery. They bridge the gap between broad dietary recommendations informed by genetic and food interactions and scientific knowledge and personalized dietary recommendations based on a person's DNA. There are four primary delivery methods to choose from. In the first model, consumers can buy and use nutrigenomics products. You can find a nutrigenomics company online, buy an online test kit, swab your cheek and send it back, and then use your results to improve your eating habits. Because of the consumer paradigm, nutrigenomics is now available to everyone. Consumers benefit from the convenience, empowerment, and foresight that comes with monitoring and forecasting a portion of their lives from the privacy and convenience of their own homes (Deb et al., 2014). Even though these tests are performed without the assistance of a medical professional, the issue of standardization and regulation of testing methodology is raised. The second approach is based on health care practitioners who provide testing and consultation, such as geneticists, primary care physicians, and nutritionists.

According to Castle, under this paradigm, patients will have access to information that is integrated into the rest of their care. Unfortunately, due to a shortage of qualified professionals, this level of care cannot be provided.

"Food and genetics are still not high on the medical training curricula," according to Castle, so doctors may be unable or unwilling to take on such a role. This section contains three models. Nutritional genomics specialists can help doctors and other health care providers. A doctor, for example, could collaborate with a company to obtain nutrigenomics reports for his or her patients while also advising and training the medical staff.

A common feature of all three models is that they all provide personalized nutrigenomics services. The emphasis in all three models is on the individual's benefits and drawbacks, whether they come from the consumer, the health care provider, or a combination of the two. All of them necessitate the use of nutrigenomics technology on a single person. Finally, the public health paradigm employs nutrigenomics to provide generalized nutrition advice to the general public. It foregoes some of the benefits of personalized dietary advice to assist a large number of people.

The United States has placed a high priority on models that provide individualized nutrigenomics counseling, but the question of whether these models are best for service delivery raises several ethical, political, and societal concerns about health care delivery. The Department of Health in the United Kingdom has produced Choosing Health: Making Healthy Choices Easier. Castle was aware of the discussion. By increasing government involvement in information dissemination and creating social structures that empower people to make their own health-related decisions, sickness prevention can be made more appealing to the public (Dang et al., 2014). The goal is to empower people to take control of their own lives and move away from a state-dominated system. Individuals in the United States are equipped with nutrigenomics technology to help them make better nutritional and health decisions.

The Food Ethics Council, a UK-based volunteer organization, was inspired to conduct its study after reading Choosing Health: Making Healthy Choices Easier. According to the report, personalization is "just one minor cog in a larger neoconservative agenda to minimize money and limit government participation in people's lives." Instead of empowering individuals, the Food Ethics Council sees customization as a way to alienate people from their governments and relieve the government of its obligation to provide health care to the general population. Rather than focusing on individual preferences, food and health should be approached from a rights-based perspective. Individuals should be guaranteed the right to a healthy diet under this model, and the government is responsible for ensuring that this right is respected. "They strongly disagree with the presentation of personal choice as the primary barrier to more public health," Castle writes. As a result, they see this as the primary impediment to improving public health.

The Food Ethics Council's arguments are critical to the future development of nutrigenomics in the United States. To begin, nutrigenomics' user base will be skewed due to the high initial cost of the technology. As a result, the vast majority of people will be unable to use technology. Castle discovered that early adopters are more likely to be educated. They have a higher level of education and are significantly wealthier than the general population. There will be a distinct divide between those who can afford the time and resources to learn about nutrigenomics and those who can't. This, according to Castle, is a major cause for concern. "The socioeconomic barrier associated with the knowledge key that is applied to acquire access into the nutrigenomics room is very considerable." Personalization agendas, if the state has a role in them, may divert attention away from the state's responsibility for public health initiatives, according to the Food Ethics Council (Dall'Asta et al., 2022). For this reason, personalization may have a role for the state to play. Castle believes that the Food Ethics Council's reasons, regardless of whether or not they are adopted, offer a clear account of the probable issues associated with the tailored paradigm for the delivery of nutrigenomics products.

While nutrigenomics may be most powerful when it comes to providing individualized recommendations to individuals, focusing on public health runs the risk of diminishing this potential. The ultimate goal of interventions like nutrigenomics is for people to live longer, happier, and healthier lives. To achieve this goal, the most efficient method is to first provide tailored counseling and then analyze the impact of this counseling on the health and well-being of the general population. Nutrigenomics health guidance cannot be evaluated effectively if it is watered down, perhaps through a public health approach. As a result, wider dissemination of the findings will be difficult. This is a question about how best to communicate nutrigenomics information to the general public in a way that they can understand and put into practice. An overabundance of information can lead to a lack of compliance with dietary guidelines, which in turn can lead to confusion. Because no one has done systematic public consultation and engagement on nutrigenomics to find out the motivations for applied nutrigenomics research as well as how people form their perceptions and the role the media plays in molding those beliefs, Castle believes this is a problem.

Not to mention, Castle speculated on whether or not nutritional genomics could be used in both individualized healthcare and population-wide health care in the future. This strategy is often recommended as a middle ground that provides the best of both worlds. If nutrigenomics is to live up to its promise of improving the lives and health of people around the world, this issue must be addressed.

10.4 SCIENCE JOURNALISM AND THE NUTRIGENOMICS REVOLUTION

In the case of nutrigenomics, basic researchers may not spend a lot of time thinking about how to effectively communicate their findings to the general public. Ultimately, nutrigenomics data will be assimilated by members of the general public as well as financial and legal authorities.

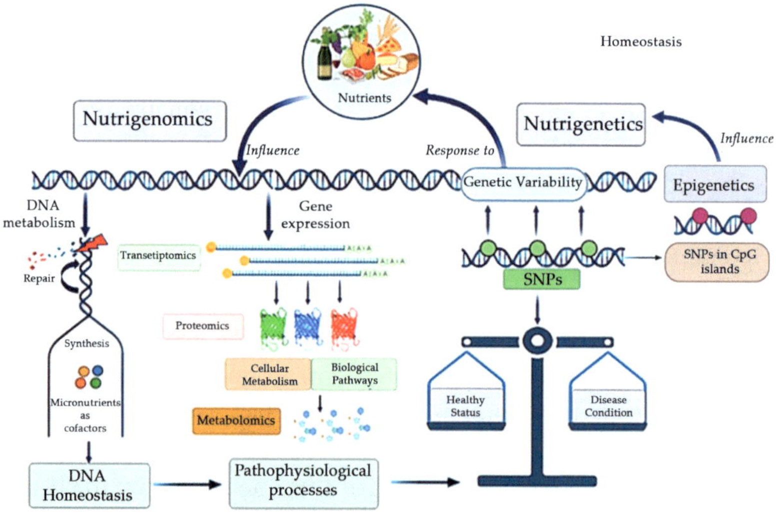

Figure 10.5: The nutrigenomics revolution.

Source: Dallio M, Romeo M, Gravina AG, Masarone M, Larussa T, Abenavoli L, Persico M, Loguercio C, Federico A. Nutrigenomics and Nutrigenetics in Metabolic- (Dysfunction) Associated Fatty Liver Disease: Novel Insights and Future Perspectives. Nutrients. 2021; 13(5):1679. https://doi.org/10.3390/nu13051679

One of the advantages of nutrigenomics is that it doesn't require a lot of effort to get the word out to the general public. Students were instructed by Sally Squires that "the public is immensely hungry for this knowledge."

"They've never been so hungry." The field, on the other hand, is at risk because of that hunger. There will be a tendency to produce results, even if they are dubious because the audience is so anxious for them, but this should be resisted. To avoid overselling and underdelivering in this region, According to Susan Okie, "it will be important not to oversell this area". She added, "the worst thing that this field could do right now is trying to market it to journalists or the public on the assumption that this area will soon come up with individualized dietary regimens for everyone based on their genes," (Elsamanoudy et al., 2016). An effective strategy for educating the general public on the benefits of nutrigenomics will be to describe it as simply and plainly as possible. A "very basic message" is required, according to Squires. You should study how politicians use sound bites to get their ideas across if you want to get your message across. As a result, you're going to need strong spokespersons who can convey a clear message, and you're going to have to deal with a lot of media that isn't highly informed in either science or medicine."

The underlying intricacy of the nutrigenomics issue will make this approach far more challenging. On this point, Okie made it clear: The idea that genes can influence how people respond to their dietary needs is complicated and highly sensitive. Okie cited research showing that people's responses to dietary changes vary widely, such as how a change in diet can raise blood pressure in some people while not affecting it in others. How do you create universal principles for the public that everyone can follow when there are so many variations in the way each person thinks? In addition, there is no guarantee that the public will follow the advice.

Concerning the two delivery models Castle spoke about, Okie pointed out that it has been extremely difficult to get people to change their personal choices despite overwhelming evidence that they can improve their chances of health by following a particular course of action. A person's family history is a form of genetic profile. They may have relatives who are overweight or diabetic or a father who suffered a heart attack at the age of 50. It was her question: Is there any evidence that people who have a genetic profile like this one are more likely to follow healthy eating guidelines, regardless of their health status? (Dimitrov et al., 2016). Despite public awareness and concern about the health risks of obesity, as well as the stigma that now exists against obese people, most people are not very successful at avoiding weight gain and are very unsuccessful, even if they manage to lose it. When conveying nutrigenomics knowledge, it appears that making it clear that people can use this information and change their lifestyles is likely to be the most difficult part of the process.

10.5 NEEDS AND OPPORTUNITIES IN THE BIOMEDICAL SCIENCES: INTERACTING NETWORKS

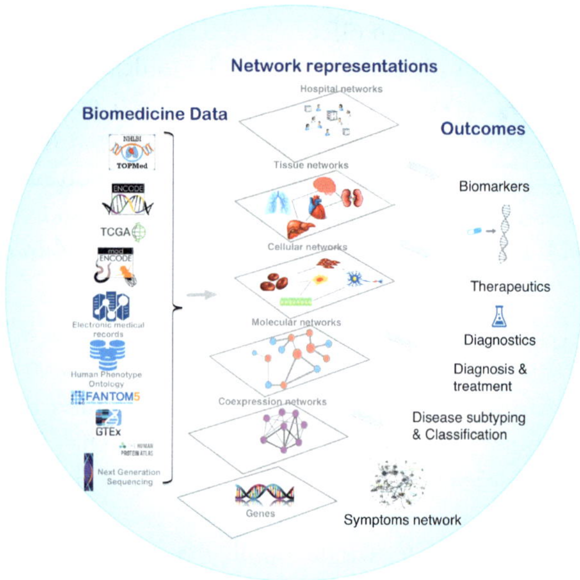

Figure 10.6: Biomedical sciences.

Source: https://www.frontiersin.org/files/MyHome%20Article%20Library/445334/445334_Thumb_400.jpg

Ralph Greenspan discusses the importance of foundational research for the development of any scientific topic, but particularly for the development of an emerging field such as nutrigenomics in this video. The scientific waste dump is quickly becoming overrun with medical models that have been successfully treated in the laboratory but have not been tried on humans. This kind of situation arises as a consequence of the many different ways in which biological networks function. It is of little consequence if sickness in an animal model and an illness in a human patient have overt or genuine clinical parallels.

As a direct consequence of this, a deeper degree of comprehension is required. The findings of the study are not impacted in any way by the question of whether or not diabetes may develop in model animals in the same manner that it can occur in people. To be able to recognize similarities across different human systems, you need to have a comprehensive grasp of operational principles. Since at least half of Drosophila's genes are also present in humans, this insect, often known as the fruit fly, is frequently used in scientific research as a model organism (Dimitrov, 2011). Even though neither genome is very large, the Drosophila genome contains roughly two-thirds of the human genes that are known to cause illness or increase the risk of disease (around 14,000 genes for Drosophila and 20,000 to 30,000 genes for humans). Because there are so few genes, it is essential to have a solid understanding of how they communicate with one another and how various gene variants might be coupled to generate certain outcomes. In conclusion, even very slight shifts in the activity of the network might have a major effect on the findings. For us to obtain a comprehensive understanding of how biological systems function, we need to do the necessary basic research. Only then will we be able to comprehend how biological activities emerge from the interactions of different genes.

10.6 GENE ORGANIZATION SYSTEMS

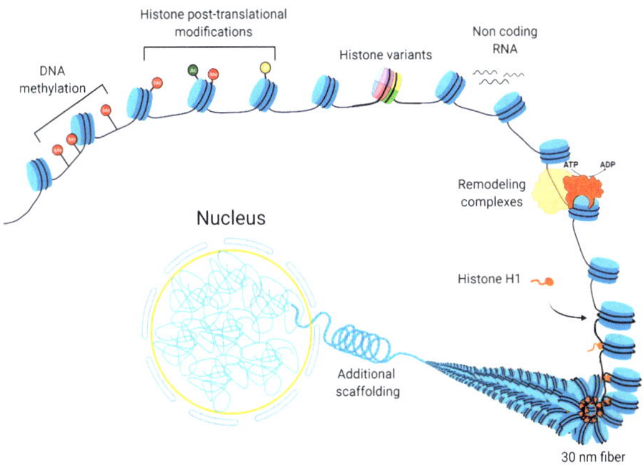

Figure 10.7: Gene Organization Systems.

Source: Alpsoy A, Sood S, Dykhuizen EC. At the Crossroad of Gene Regulation and Genome Organization: Potential Roles for ATP-Dependent Chromatin Remodelers in the Regulation of CTCF-Mediated 3D Architecture. Biology. 2021; 10(4):272. https://doi.org/10.3390/biology10040272

Because of a mutation in the syntaxin gene, Drosophila cannot survive temperatures that are higher than 39 degrees Celsius when this mutation is present (102 degrees Fahrenheit). Flies that do not have the mutation are far more difficult to kill with heat (the flies finally get clumsy and fall to the bottom of a glass tube that contains them) (normal flies without the mutation are also affected by these temperatures, and they eventually become uncoordinated and fall to the bottom of a glass tube holding them, but it takes far longer for the heat to takes its toll). Flies were found to be able to remain active even when exposed to high temperatures for significantly longer periods thanks to the discovery of sixteen additional mutations that decreased this tendency (Wiwanitkit, 2012). If a fly had not just the syntaxin 1A gene mutation but also another mutation, it would be able to tolerate higher temperatures than a fly that had only the syntaxin 1A gene mutation.

The ability of the flies to withstand heat was evaluated after they were reared without the syntaxin 1A mutation. They were able to determine how the 16 different mutations interacted with one another by using this method. A mathematical estimate of how each pairing would affect a fly's ability to withstand heat in the absence of any interaction between the individual pairs of genes was generated by using the data for the various pairings. This estimate was created in the absence of any interaction between the individual pairs of genes. All of the impacts would add up to one total if the individual gene pairs did not interact with one another. When the observed and anticipated results were compared to determine whether or not certain combinations of genes had synergistic interactions, it was discovered that certain combinations do have beneficial effects. If the impacts of these genes were just summed up, the cumulative effect on the fly's heat tolerance would not have been nearly as significant as it was when they were combined (Wittwer et al., 2011). On the other hand, several gene combinations displayed unfavorable interactions with one another. The synergistic effects of any other combinations were completely nonexistent. In the wake of the finding of a gene network, a further experiment studied how the network develops in response to a genetic mutation in a particular fly locus, namely the gene responsible for the production of syntaxin 1A. As a direct consequence of this fact, the synergistic interactions that take place between genes can be practically entirely distinct depending on the genotype that is present at the syntaxin 1A locus.

This particular kind of gene network is not stable; rather, it is dependent on the particular mutations of the genes that make up the network. These mutations can cause the network to change. When one gene in a network changes, the changes have the potential to affect a large number of other genes in the network, which in turn can have a considerable impact on the network as a whole. Any adjustments that you make to a network will have repercussions farther down the line. For example, in the case of syntaxin 1A, it was previously thought that split genes did not influence the generation of synaptic vesicles in neurons, which is one of the syntaxin's roles (which is where syntaxin has a role). The genes that code for proteins that are essential to the metabolism and the cytoskeleton have been altered in a significant number of instances. Networks that have several connections are referred to as "multiplexes."

10.7 NEEDS AND OPPORTUNITIES IN THE FOOD AND AGRICULTURAL SCIENCES

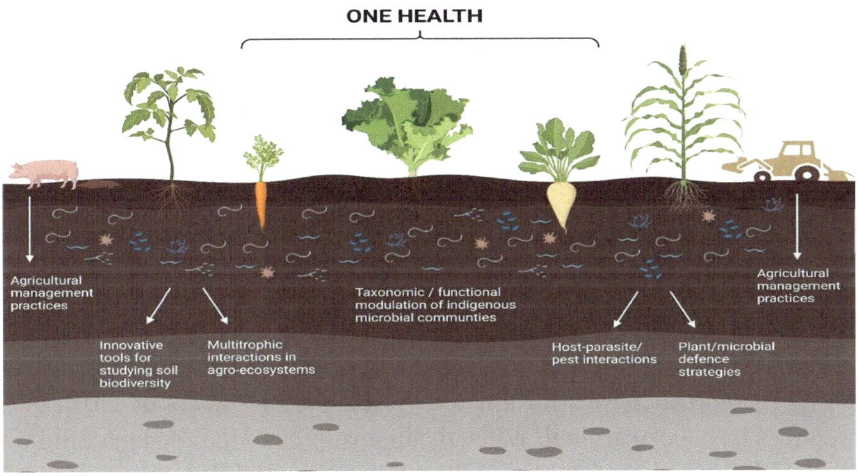

Figure 10.8: Agricultural Sciences.

Source: https://www.mdpi.com/journal/sustainability/special_issues/soil_microbiomes_agriculture

Nutrition genomics has the potential to improve our knowledge of nutrition and motivate people to eat healthier, but this promise will not be fulfilled until equivalent progress is achieved in other areas of study. Two of these sectors are agriculture and the food business. Agriculture and the food sector, according to Joseph Spence, can help nutrigenomics realize its full potential.

Among other things, Spence projected that this role would evolve in several ways, the most significant of which would be the change of the nation's food supply to reflect the increasing nutritional requirements, which is now being done on an experimental basis. The NuSun sunflower, created by the Agricultural Research Service (ARS), has a high concentration of monounsaturated fatty acids and oleic acid. Approximately 77% of the sunflowers grown in the United States for their oil seeds come from a single variety. The NuSun sunflower was developed using normal breeding methods, but genetic engineering was also applied to create desired varieties (Van Breda et al., 2015). Transgenic tomatoes with four to eight times the usual amount of lycopene have also been developed by ARS researchers. Known for its potent antioxidant properties, lycopene is a carotenoid known as lycopene.

Nutrigenomics research is generating more knowledge on the impacts of different nutrients; hence, the agricultural sector may change food production to incorporate these findings into future research and development as the area of nutrigenomics continues to progress. It will be critical to have research that shows the unique nutritional benefits of these varied nutrients. "We can't make the mistake of declaring that this will almost certainly be fantastic," Spence said, which is a trap. There must be a greater emphasis on the long-term health benefits, as well as a greater emphasis on the immediate benefits. Additionally, the US has frequently served as a go-between in international disputes throughout history. Using tools such as the food pyramid, the United States Department of Agriculture (USDA) has been making recommendations for a healthy diet since 1992. As of 2010, these guidelines have been around. When the first version of the food pyramid was developed, it was based on a population-based model that was universally applicable. When it comes to the current version (MyPyramid), numerous pyramids are based on an individual's age, gender, and level of physical exertion. Many variables influence a person's nutritional requirements.

The field of nutritional genomics will allow us to build on the current trend of nutritional counseling and provide even more personalized dietary recommendations in the years to come. It is "incredibly difficult" to continue to provide dietary advice based on population-based studies "given our various genetic origins, lifestyles, etc.," as Spence put it. In the not-too-distant future, nutritional advice will take into account an individual's genetic background and be adjusted by the family tree (Su & Lu, 2012). Nutrigenomics studies are likely to identify ways in which nutritional advice should differ between, for example, American Indian/Alaska Natives and descendants of northern Europeans; it is recommended that nutritional guidelines be revised to reflect these insights. However, relatively little research has been done in this field to this point.

10.8 THE FUTURE ECONOMIC POTENTIAL OF NUTRIGENOMICS

The possibility of making more money is frequently cited as one of the primary motivations behind the creation of new technologies. It is more likely that a technology that has outstanding financial prospects would attract more financing and make more strides toward advancement than other technologies whose financial prospects are less certain. During her presentation, Patricia Danzon discussed the potential financial influence that nutrigenomics may have on the decrease of the costs associated with medical care.

Because the field of nutrigenomics is still in its infancy, Danzon believes that attempting to evaluate the economic repercussions at this time would be a reckless act. Before drawing to a close with her presentation, she shifted her focus to the monetary potential of nutrigenomics and the methods that may be used to its full advantage. The great majority of monies allocated to health care are utilized for the treatment of patients as opposed to the prevention of disease. Danzon made this claim about Tommy Thompson, who served as the former Secretary of Health and Human Services. He stated that 95 percent of the money spent on health care goes toward treating patients who already have an illness, while only 5 percent is spent on preventing illnesses in healthy people. It's possible that increasing one's investment in preventative measures that are shown to work might result in a big economic payback.

According to the opinions expressed by the industry experts, nutrigenomics can exert a significant impact on the economic climate of this particular sector. Danzon provided examples of two separate occurrences to back up his allegation. It has not been determined whether or not individuals are willing to change their routines in response to advice based on nutrigenomics (Singh & Sharma, 2019). When it comes to the potential of a certain field, its actualization does not always require the same approach. You state, "I am surprised by the fact that so little has been developed [in the pharmaceutical sector] during the previous five years in comparison to what was promised by the more hopeful folks around the time that the human genome was mapped," and you are right to be surprised by this. "I am surprised by the fact that so little has been developed [in the pharmaceutical sector] during the previous five years." I respond by saying, "This takes me by surprise." Despite this, nutrigenomics has the potential to bring enormous financial rewards; however, this can only occur if significant advancements are made in the science that underpins it.

It is anticipated that the economic benefits of nutrigenomics will continue for an extended length of time. Altering one's food, on the other hand, is more likely to have long-term impacts, which may have an influence on one's health in the years to come. As a consequence of this, nutrigenomics has the potential to have a major influence on the severity of an illness or the course of a disease. It is more probable that disease prevention initiatives will be successful and cost-effective if they center their attention on high-risk illnesses and the demographic groups who are most likely to be affected by such illnesses. Cost savings from preventative procedures are only achieved when a disease is averted in those who would have otherwise contracted it. Preventive procedures are expensive. This is because although everyone is a target of the preventative effort, only those people who would have developed the condition are rescued. As a consequence of this, disease prevention is more successful when efforts are taken to avoid diseases that are particularly expensive to treat. As a direct consequence of this, persons who belong to high-risk groups are more likely to become unwell.

It is feasible to reduce the expenses of preventative care, which would make the practice more cost-effective if it were done. According to Danzon, it is more cost-effective to adjust meals to achieve a balanced diet than it is to supplement an already existing diet with costly vitamins and minerals. She bemoaned the fact that the current system encourages individuals to use supplements rather than pushing them to have a diet that includes a wide variety of foods. Although insurance companies may be required to pay for medications or supplements that are deemed to be medically essential, they are not often required to pay for better diets, even though such diet changes are frequently recommended by medical professionals. In the same vein, the decisions that consumers make may be influenced if the FDA gives its blessing to dietary supplements but remains silent on the question of therapeutic foods (Sikalidis, 2019). A new era will require a shift in how we think about insurance coverage, and new rules will be essential. Both of these things are going to be necessary. We want to make sure that we promote this trend so that we can avoid distorting it in any way, and we want to make sure that we do so."

In his final argument, Danzon underlined the significance of convincingly showing the usefulness of the many nutrigenomics treatments that are already available. If the signal of the nutrigenomics data is consistently obscured by the "noise" that customers are constantly exposed to, then fewer people will pay attention to the nutrigenomics data. Because of this, the signal-to-noise ratio must be high enough to affect the behavior of the client (Riscuta, 2016). According to Danzon, "nutrigenomics research has the potential to have significant economic advantages if that can be done, and if care can be made to target high-risk illnesses and populations, and to rethink how regulatory and insurance systems work."

REFERENCES

1. Alagawany, M., Elnesr, S. S., Farag, M. R., El-Naggar, K., & Madkour, M. (2022). Nutrigenomics and nutrigenetics in poultry nutrition: An updated review. *World's Poultry Science Journal*, *78*(2), 377-396.
2. Afman, L., & Müller, M. (2006). Nutrigenomics: from molecular nutrition to prevention of disease. *Journal of the American Dietetic Association*, *106*(4), 569-576.
3. Banerjee, G., Pal, R., & Ray, A. K. (2015). Applications of nutrigenomics in animal sectors: A Review. *Asian Journal of Animal and Veterinary Advances*, *10*(9), 489-499.
4. Benítez, R., Núñez, Y., Óvilo, C., & Ovilo, C. (2017). Nutrigenomics in farm animals. *J Invest Genomics*, *4*, 1.
5. Bragazzi, N. L. (2013). Situating nutri-ethics at the junction of nutrigenomics and nutriproteomics in postgenomics medicine. *Current Pharmacogenomics and Personalized Medicine (Formerly Current Pharmacogenomics)*, *11*(2), 162-166.
6. Braicu, C., Mehterov, N., Vladimirov, B., Sarafian, V., Nabavi, S. M., Atanasov, A. G., & Berindan-Neagoe, I. (2017, October). Nutrigenomics in cancer: Revisiting the effects of natural compounds. In *Seminars in Cancer Biology* (Vol. 46, pp. 84-106). Academic Press.
7. Brennan, L., & de Roos, B. (2021). Nutrigenomics: lessons learned and future perspectives. *The American Journal of Clinical Nutrition*, *113*(3), 503-516.

8. Bionaz, M., Osorio, J., & Loor, J. J. (2015). TRIENNIAL LACTATION SYMPOSIUM: Nutrigenomics in dairy cows: Nutrients, transcription factors, and techniques. *Journal of animal science*, *93*(12), 5531-5553.
9. Castro, A. G., Rocca-Serra, P., Stevens, R., Taylor, C., Nashar, K., Ragan, M. A., & Sansone, S. A. (2006). The use of concept maps during knowledge elicitation in ontology development processes–the nutrigenomics use case. *BMC bioinformatics*, *7*(1), 1-14.
10. Carlberg, C. (2019). Nutrigenomics of vitamin D. *Nutrients*, *11*(3), 676.
11. Chen, L., & Zhao, E. M. (2013). Advances of nutrigenomics: concept, content, technology and development. *Journal of Experimental Biology*, *1*, 3.
12. Corella, D., & Ordovas, J. M. (2009). Nutrigenomics in cardiovascular medicine. *Circulation: Cardiovascular Genetics*, *2*(6), 637-651.
13. Connolly, A. J., & Dawson, K. A. (2015). A glimpse into the concept of nutrigenomics. *AFMA Matrix*, *24*(1), 41-45.
14. Deb, R., Chakraborty, S., Verma, A. K., Tiwari, R., & Dhama, K. (2014). Nutrigenomics and its role in male puberty of cattle: a mini review. *Pakistan Journal of Biological Sciences: PJBS*, *17*(3), 329-334.
15. Dang, T. S., Walker, M., Ford, D., & Valentine, R. A. (2014). Nutrigenomics: the role of nutrients in gene expression. *Periodontology 2000*, *64*(1), 154-160.
16. Dall'Asta, M., Barbato, M., Rocchetti, G., Rossi, F., Lucini, L., Marsan, P. A., & Colli, L. (2022). Nutrigenomics: an underestimated contribution to the functional role of polyphenols. *Current Opinion in Food Science*, 100880.
17. Dimitrov, D. V. (2011). The human gutome: nutrigenomics of the host–microbiome interactions. *OMICS: A Journal of Integrative Biology*, *15*(7-8), 419-430.
18. Dimitrov, D., Thiele, I., & Ferguson, L. R. (2016). The Human Gutome: Nutrigenomics of Host-Microbiome Interactions. *Frontiers in Genetics*, *7*, 158.
19. Elsamanoudy, A. Z., Neamat-Allah, M. A. M., Mohammad, F. A. H., Hassanien, M., & Nada, H. A. (2016). The role of nutrition related genes and nutrigenetics in understanding the pathogenesis of cancer. *Journal of Microscopy and Ultrastructure*, *4*(3), 115-122.

20. Fallaize, R., Macready, A. L., Butler, L. T., Ellis, J. A., & Lovegrove, J. A. (2013). An insight into the public acceptance of nutrigenomic-based personalised nutrition. *Nutrition research reviews*, *26*(1), 39-48.
21. Fenech, M. (2008). Genome health nutrigenomics and nutrigenetics–diagnosis and nutritional treatment of genome damage on an individual basis. *Food and Chemical Toxicology*, *46*(4), 1365-1370.
22. Ferguson, L. R. (2016). The value of nutrigenomics science. *OMICS: A Journal of Integrative Biology*, *20*(2), 122-122.
23. Ferguson, L. R., Philpott, M., & Barnett, M. P. (2010). Nutrigenomics: integrating genomic approaches into nutrition research. *Molecular Diagnostics*, 347-363.
24. Fenech, M., El-Sohemy, A., Cahill, L., Ferguson, L. R., French, T. A. C., Tai, E. S., ...& Head, R. (2011). Nutrigenetics and nutrigenomics: viewpoints on the current status and applications in nutrition research and practice. *Lifestyle Genomics*, *4*(2), 69-89.
25. Guéant, J. L., Elakoum, R., Ziegler, O., Coelho, D., Feigerlova, E., Daval, J. L., & Guéant-Rodriguez, R. M. (2014). Nutritional models of foetal programming and nutrigenomic and epigenomic dysregulations of fatty acid metabolism in the liver and heart. *Pflügers Archiv-European Journal of Physiology*, *466*(5), 833-850.
26. Gonzalez, M. J., Miranda-Massari, J. R., Duconge, J., Rodriguez, J. R., Cintron, K., Berdiel, M. J., & Rodriguez, J. W. (2015). Nutrigenomics, Metabolic Correction and Disease. *Journal of Restorative Medicine*, *4*(1), 74-82.
27. Hesketh, J. (2013). Personalised nutrition: how far has nutrigenomics progressed? *European journal of clinical nutrition*, *67*(5), 430-435.
28. Hurlimann, T., Menuz, V., Graham, J., Robitaille, J., Vohl, M. C., & Godard, B. (2014). Risks of nutrigenomics and nutrigenetics? What the scientists say. *Genes & nutrition*, *9*(1), 1-12.
29. Irimie, A. I., Braicu, C., Pasca, S., Magdo, L., Gulei, D., Cojocneanu, R., ...& Berindan-Neagoe, I. (2019). Role of key micronutrients from nutrigenetic and nutrigenomic perspectives in cancer prevention. *Medicina*, *55*(6), 283.
30. Kang, J. X. (2012). Nutrigenomics and systems biology. *Lifestyle Genomics*, *5*(6), I-II.

31. Kussmann, M., & Van Bladeren, P. J. (2011). The extended nutrigenomics–understanding the interplay between the genomes of food, gut microbes, and human host. *Frontiers in genetics*, *2*, 21.
32. Kussmann, M., & Affolter, M. (2009). Proteomics at the center of nutrigenomics: comprehensive molecular understanding of dietary health effects. *Nutrition*, *25*(11-12), 1085-1093.
33. Ladeira, M. M., Schoonmaker, J. P., Swanson, K. C., Duckett, S. K., Gionbelli, M. P., Rodrigues, L. M., & Teixeira, P. D. (2018). Nutrigenomics of marbling and fatty acid profile in ruminant meat. *Animal*, *12*(s2), s282-s294.
34. Loor, J. J., Vailati-Riboni, M., McCann, J. C., Zhou, Z., & Bionaz, M. (2015). Triennial Lactation Symposium: Nutrigenomics in livestock: Systems biology meets nutrition. *Journal of animal science*, *93*(12), 5554-5574.
35. Mathers, J. C. (2017). Nutrigenomics in the modern era. *Proceedings of the Nutrition Society*, *76*(3), 265-275.
36. Masotti, A., Da Sacco, L., Bottazzo, G. F., & Alisi, A. (2010). Microarray technology: a promising tool in nutrigenomics. *Critical reviews in food science and nutrition*, *50*(7), 693-698.
37. Mustapa, M. A. C., Amin, L., & Frewer, L. J. (2020). Predictors of stakeholders' intention to adopt nutrigenomics. *Genes & nutrition*, *15*(1), 1-15.
38. McNamara, J. P. (2010). *Nutrigenomics for improved reproduction* (pp. 413-435). Wiley-Blackwell, Iowa.
39. Neeha, V. S., & Kinth, P. (2013). Nutrigenomics research: a review. *Journal of food science and technology*, *50*(3), 415-428.
40. Pal, A. (2018). Nutrigenomics-an emerging area. *Acta Scientific Nutritional Health*, *2*(9), 12-13.
41. Piroddi, M., Albini, A., Fabiani, R., Giovannelli, L., Luceri, C., Natella, F., ...& Galli, F. (2017). Nutrigenomics of extra-virgin olive oil: A review. *Biofactors*, *43*(1), 17-41.
42. Peña-Romero, A. C., Navas-Carrillo, D., Marín, F., & Orenes-Piñero, E. (2018). The future of nutrition: Nutrigenomics and nutrigenetics in obesity and cardiovascular diseases. *Critical reviews in food science and nutrition*, *58*(17), 3030-3041.

43. Pauletto, M., Elgendy, R., Ianni, A., Marone, E., Giantin, M., Grotta, L., ...& Martino, G. (2020). Nutrigenomic effects of long-term grape pomace supplementation in dairy cows. *Animals*, *10*(4), 714.

44. Qadeer, A., & Ashhar, M. U. (2014). Concept of Nutrigenomics Exists in Unani Medical Philosophy. *RADS Journal of Pharmacy and Pharmaceutical Sciences*, *2*(2), 26-30.

45. Ray, S. (2014). Micronutrient, genome stability and degenerative diseases: nutrigenomics concept of disease prevention–an overview. *Current Research in Nutrition and Food Science Journal*, *2*(3), 159-164.

46. Riscuta, G., & Dumitrescu, R. G. (2012). Nutrigenomics: implications for breast and colon cancer prevention. *Cancer Epigenetics*, 343-358.

47. Riscuta, G. (2016). Nutrigenomics at the interface of aging, lifespan, and cancer prevention. *The Journal of Nutrition*, *146*(10), 1931-1939.

48. Sikalidis, A. K. (2019). From food for survival to food for personalized optimal health: A historical perspective of how food and nutrition gave rise to nutrigenomics. *Journal of the American College of Nutrition*, *38*(1), 84-95.

49. Singh, R., & Sharma, L. (2019). Nutrigenomics: A combination of nutrition and genomics: A new concept. *J Physiol Nutr Phys Ed*, *4*, 417-421.

50. Su, H. L., & Lu, T. J. (2012, July). Exploring the consumer acceptance of and preferences in nutrigenomics-based personalized health management service. In *2012 Proceedings of PICMET'12: Technology Management for Emerging Technologies* (pp. 3050-3058). IEEE.

51. Van Breda, S. G., Wilms, L. C., Gaj, S., Jennen, D. G., Briedé, J. J., Kleinjans, J. C., & De Kok, T. M. (2015). The exposome concept in a human nutrigenomics study: evaluating the impact of exposure to a complex mixture of phytochemicals using transcriptomics signatures. *Mutagenesis*, *30*(6), 723-731.

52. Wittwer, J., Rubio-Aliaga, I., Hoeft, B., Bendik, I., Weber, P., & Daniel, H. (2011). Nutrigenomics in human intervention studies: current status, lessons learned and future perspectives. *Molecular nutrition & food research*, *55*(3), 341-358.

53. Wiwanitkit, V. (2012). Database and tools for nutrigenomics: A brief summary. *Journal of Medical Nutrition and Nutraceuticals*, *1*(2), 87.

INDEX

Printed in the United States
by Baker & Taylor Publisher Services